THE FRONTIERS COLLECTION

THE FRONTIERS COLLECTION

Series Editors

A. C. Elitzur   Z. Merali   T. Padmanabhan   M. Schlosshauer
M. P. Silverman   J. A. Tuszynski   R. Vaas

The books in this collection are devoted to challenging and open problems at the forefront of modern science, including related philosophical debates. In contrast to typical research monographs, however, they strive to present their topics in a manner accessible also to scientifically literate non-specialists wishing to gain insight into the deeper implications and fascinating questions involved. Taken as a whole, the series reflects the need for a fundamental and interdisciplinary approach to modern science. Furthermore, it is intended to encourage active scientists in all areas to ponder over important and perhaps controversial issues beyond their own speciality. Extending from quantum physics and relativity to entropy, consciousness and complex systems the Frontiers Collection will inspire readers to push back the frontiers of their own knowledge.

More information about this series at http://www.springer.com/series/5342

For a full list of published titles, please see back of book or springer.com/series/5342

Sergio Carrà

Stepping Stones to Synthetic Biology

 Springer

Sergio Carrà
Department of Chemistry,
Materials and Chemical Engineering
Polytechnic University
Milan, Italy

ISSN 1612-3018   ISSN 2197-6619   (electronic)
THE FRONTIERS COLLECTION
ISBN 978-3-030-07041-0   ISBN 978-3-319-95459-2   (eBook)
https://doi.org/10.1007/978-3-319-95459-2

This Springer imprint is published by the registered company Springer Nature Switzerland AG
The registered company address is: Gewerbestrasse 11, 6330 Cham, Switzerland

Preface

The sentence *"What I cannot create, I do not understand"*, left on his blackboard by Richard Feynman, echoes the formula *"Verum ipsum factum"* used by the Italian philosopher Giambattista Vico to point to the guideline of his new science (Scienza Nuova, 1725) in which man can truly know that which is produced and executed by him. In fact, only in this way can he know the exact genesis of things in the world. This message has been reiterated by the American biotechnologist and businessman Craig Venter, one of the pioneers of the human genome decryption, who asserts: *"What I cannot build, I cannot understand"*. With this message, he leads us to the extraordinary scientific adventures of the creation of an artificial cell. Venter and his colleagues have been engaged in investigations into the production of a synthetic cell, the simplest among the natural ones. Up to now, they have been able to synthesize specific oligonucleotide molecules that can reproduce themselves after being inserted into a suitable cell deprived of its natural genetic material.

The result represents a significant step in a research programme aimed at understanding and, in some respects, controlling the life processes. It includes several research areas that encompass microbiology, physical chemistry and information theory. The ongoing developments appear relevant, since they offer an engineering approach to genetics through the perspective of creating life itself, thanks to the significant results obtained in the control of some networks of chemical reactions occurring in the cells and in the processes involved in cellular transmission of matter and energy.

Actually, life should not be thought of only as a chemical event, but also as a process involving information transmission, because the genome is a repository of information gathered over time through evolution. Application of the above concepts to microbiological growth is improving knowledge about the mechanisms that affect the increase in complexity associated with evolutionary paths. Moreover, it suggests analogies with the evolution of the technologies connected with the development of our society. Finally, it suggests approaches for the management of energy transformations occurring in large-scale activities.

The aforementioned topics, as well as being characterized by scientific and philosophical content, are inspiring ongoing research baptized as synthetic biology and are giving impetus to the employment of bacterial cells as factories for the conversion of renewable resources to chemical products, whose applications range from pharmaceuticals to biofuels. In fact, the creation of new reaction pathways constitutes the core of a new field called metabolic engineering, whose focus is the activation of biological functions nonexistent in nature. On the whole, it represents a new approach to industrial production activities coming out of the synergic combination of biology, chemistry, information theory and engineering.

Such topics at the edge of life will be explored, starting from a topical subject from the discussions in which science and philosophy merge, that is, the connection of life processes with the second law of thermodynamics, which, according to Sir Arthur Eddington, *"holds the supreme position among the laws of Nature"*. Then, we move on to the investigation into molecular mechanisms, whereby order is formed out of chaos.

Milan, Italy   Sergio Carrà

Contents

Employed Symbols

C_i Concentration of component i (*moles per volume*)

G Free energy

H Enthalpy

I Information

J Flux

m Molecular mass

M Mass

n_i Moles number of the component i

n_{tot} Total number of moles

N Number of molecules

P Pressure

r_k Rate of k-th reaction

R_i Rate of formation of component i

S Entropy

t Time

T Temperature

U Internal energy

v Velocity

V Volume

x_i Mole fraction of component i ($x_i = n_i/n_{tot}$)

η Efficiency

ρ Density

μ Viscosity

μ_i Chemical potential of component i

ν Frequency

ε_r Energy of the r-th state

ν_{ki} Stoichiometric coefficient of component i in the k-th reaction

Φ Volumetric flow

Ψ Exergy

Mathematical Notation

x, y, z	Cartesian coordinates
r	Radial vector in a system of Cartesian coordinates
∇	Gradient
Δ	Change of a state function
$\nabla^2 = \frac{\partial^2}{\partial x^2} + \frac{\partial^2}{\partial y^2} + \frac{\partial^2}{\partial z^2}$	Laplacian operator
$\langle \rangle$	Medium value
[ ]	Molar concentration

Physical Constants

Light velocity	$c = 2.997925 \times 10^{10}$ cm/s
Electron charge	$e = 4.8023 \times 10^{-19}$ electrostatic units
Avogadro number	$N_A = 6.023 \times 10^{23}$ molecules/g mol
Planck constant	$h = 6.6261 \times 10^{-27}$ erg s
Perfect gas constant	$R = 1.98$ cal/mol K
Boltzmann constant	$k_B = R/N_A = 1.3807 \times 10^{-14}$ erg/K
Faraday constant	$F = 96485.3365$ coulombs/mol

Chapter 1
Devils, Ratchets and Biomolecular Motors

1.1 The Century of the Devils

Some students like to wear T-shirts decorated with eye-catching, although cryptic, mathematical equations that disclose the existence of light radiations. This writing is attributed to God through the implicit presence of the biblical "Fiat Lux". Actually, this is plagiarism, because the actual author was James Clerk Maxwell, a British mathematician and physicist of the eighteenth century, well known for having formulated a theoretical synthesis of electricity and magnetism.

A leading scientist in the landscape of the nineteenth century, he did not overlook the opportunity to make some contributions to thermodynamics, the new scientific achievement born as a result of the problems coming from the industrial revolution then in progress, related to the performance of thermal machines. Energy being the propensity to perform work, attention was focused on the transformation of the thermal energy released by the combustion processes into the mechanical energy capable of moving the afore-mentioned machines.

The birth of thermodynamics was officially coincident with the publication in 1823 of a book by the young Napoleonic officer Sadi Carnot, with the glowing title "*Reflections on the Motive Power of Fire*". Its most important achievement was proof of the existence of a threshold at the transformation of heat into mechanical work.

A few years later, in a significant publication of 1854, Rudolf Clausius, born in Germany but, at that time, professor at ETH in Zurich, lent thermodynamics a new degree of relevance that went beyond the study of the processes taking place in heat engines, because it extended those processes to all transformations occurring in nature, including those related to living organisms. The validity of the new formulation lay in the identification of a physical quantity that he called entropy S, from the ancient Greek words ἐν (*en*) "in", and τροπή (*tropé*) "transformation", meaning 'upheaval', which would later play an important role in scientific and philosophical culture, even in regard to its mediatic aspects. Operatively, it is expressed by the ratio

© Springer Nature Switzerland AG 2018
S. Carrà, *Stepping Stones to Synthetic Biology*, The Frontiers Collection,
https://doi.org/10.1007/978-3-319-95459-2_1

$S = (energy/T)$ between the total energy of the system and the temperature. In spontaneous processes, such as the transmission of heat between two bodies at different temperatures or the expansion of a gas subject to a pressure difference, entropy cannot decrease. This statement has been elevated to the role of the second law of thermodynamics, leading to the extension of the principle of energy conservation to include thermal energy, which has the privilege of occupying the top position.

If the view is enlarged to the entire universe, it follows that its evolution proceeds towards a final state of equilibrium, characterized by a uniform value of temperature. Lord Kelvin (William Thomson), a British physicist and engineer who made such important contributions to thermodynamics that the absolute temperature T came to be denoted by his name, defined the occurrence as "thermal death". Such a disturbing teleology was also defined by Clausius as "disintegration", thereby opening a debate that is still active and involves many aspects of scientific and philosophical culture.

Even science has sometimes been subject to the fascination human beings have with supernatural abilities, the ability to solve complex problems that are unapproachable from a human capacity, by making use of the most orthodox determinism, implying a necessary relation between causes and effects. In this context, Pierre-Simon de Laplace, one of the leading French scientists in the period between the eighteenth and nineteenth centuries, wrote a treatise on the system of the world in which he offered a description of the movements of the planets of the solar system. It is said that during the presentation of the book in the presence of the Emperor Napoleon, the author replied to Napoleon's observation that God was never mentioned by saying that it was an unnecessary hypothesis. The Emperor answered that, nevertheless, it could explain several things.

In 1814, Laplace claimed that a being endowed with a monstrous intelligence would be able accurately to foresee the future events of a physical system if informed of the minute details of its initial state. This hypothetical character entered into the culture with the nickname of the Laplace devil.

Maxwell's attention was instead focused on the significance of the second law of thermodynamics, through his study of systems whose behavior depends on random events. For instance, the chaotic movements of the molecules present in a gas that exchange energy through mutual collisions. Maxwell became involved in such a problem at the request of one of his friends, Peter Guthrie Tait, that he make a contribution to a book on Thermodynamics. In the answer, Maxwell evoked the image of a being with extraordinary faculties, because it was able to violate the second law of thermodynamics. Referring to a system formed by two containers separated by a gate valve containing the molecules of a gas subject to thermal agitation and at different temperatures, the being whose faculties are so sharpened that it can follow every molecule in its course would be able do what is impossible to us. In fact, it could be conceived as a device able to identify and separate fast (hot) and slow (cold) molecules in order to set up a temperature or pressure gradient, thus producing an asymmetrical configuration with a decrease of the entropy.

Maxwell concluded that the validity of the second law of thermodynamics is not absolute, but depends on our inability to follow the motion of individual molecules and identify their speed.

The devil would be exorcised a century later by Leo Szilard, the Hungarian physicist, who, together with Einstein, wrote the famous letter to Roosevelt to urge him in the promotion of research into nuclear energy. In a publication of 1929, Szilard showed that the acquisition of the information needed by the Maxwell devil to operate the gate involves the creation of a quantity of entropy equal to that generated in the mixing process of the molecules that were separated. Although the analysis carried out by Szilard was not entirely correct, it became the starting point of a wide variety of activity that would influence important aspects of scientific culture, with fallout for the communications industry, physics, chemistry and, particularly, biology.

1.2 The Unbearable Attractiveness of Irreversibility

The justification of the increase of entropy during natural transformations has always been, indeed still is, a thorny problem. Such transformations are called irreversible, because they can take place only in one direction, and once the final state is reached, it is impossible to go back to the starting point without leaving a trace in the world outside of the system. This peculiarity is associated with the existence of an "arrow of time", whose existence cannot be neglected in any attempts to give realistic descriptions of natural phenomena, particularly when biology is involved, because its objects are subject to lifecycles. Research into cells cannot neglect their progress from birth, which results from the division of a mother cell, to reproduction, which corresponds with the division that creates two new daughter cells.

The issue was addressed in its fundamental aspects by Ludwig Boltzmann, a leading figure in nineteenth century physics for the innovative contributions he made to thermodynamics and statistical mechanics. Significant is the following farsighted sentence that emphasizes the relevance of entropy in the solution of problems concerning biological processes:

> The general struggle for existence of animate beings is not a struggle for raw materials—these, for organisms, are air, water and soil, all abundantly available—nor for energy, which exists in plenty in any body in the form of heat, but of a struggle for entropy, which becomes available through the transition of energy from the hot sun to the cold earth.

He spent most of his life as a university professor at Graz in Austria. Despite being esteemed, the last years of his life were marred by disagreement with the philosophical-scientific establishment of the late nineteenth century, which cast doubt on the atomic theory of matter. The harsh tone of the conflict threw him into a state of depression that led to his suicide in 1906. A few years later, in 1913, the French physicist Jean-Baptiste Perrin, in a book that experienced great success, published the results of researches aimed at the interpretation of the behavior of dispersions of nanoscale particles in a liquid by using a theory published by Einstein in 1905. His achievement at last sanctioned the existence of atoms, even though, in the meantime, their essential role in the development of scientific activities had already emerged.

Boltzmann was an admirer of Darwin, so much so that he aimed to describe the evolution of physical systems, taking that of living organisms as a model. Aware of the tools that mathematical physics offered him for undertaking the task, he started by considering the simplest system of thermodynamics, that is, a gas formed by a large number of molecules. In his most important publication of 1872, he described its evolution through an approach that combined statistics, used for the description of the distribution of velocities of the molecules compared to hard spheres, with the dynamics of their collisions. A gas in which a pressure difference is present undergoes an expansion, which leads it towards a final state of equilibrium at which the pressure is uniform throughout the entire system. The impossibility of describing the motions of each molecule implied the use of a distribution function, which expresses the probability that a molecule occupies a given position in space with a determined value of speed. For a gas that is not under equilibrium conditions, such a function obviously depends on time. Technically, any further development implies the solution of a complex equation bearing the name of Boltzmann, an equation that is considered an icon of mathematical physics, although it has not yet acquired the dignity of decorating the T-shirts of the students. For those who are curious, its features are summarized in Box 1.1. Through its application, it can be shown that the spontaneous expansion of the gas is accompanied by an increase of entropy in agreement with the second law of thermodynamics, which was so demonstrated starting from the fundamental laws of mechanics.

To turn off the enthusiasm on such an amazing outcome, it must be mentioned that a corrosive criticism was raised by Josef Loschmidt, a German physical chemist, who recalled that, in the Newtonian mechanics employed by Boltzmann to describe the molecular collisions, time is reversible. In fact, the equations of motion are symmetric: they remain unchanged if the sign of the variable time is changed. Contrastingly, in thermodynamics, time is asymmetric, consistent with the monotonous increase of entropy, as well as the presence of the afore-mentioned time arrow. The objection inspired embarrassment in the scientific community, by opening up a discussion that has lasted to the present day.

A further in-depth analysis gave evidence that, paradoxically, the success of the Boltzmann equation in predicting the perennial increase of entropy is due to an approximation, called "molecular chaos", which implies the absence of any correlation between the motion of the molecules. Thus, the collisions are occurring at random.

Actually, the descriptions of the transformations occurring in the microscopic world depend on the scale of observation, which requires the subdivision of the space into cells by following a subjective method in the choice of their size. The specification of the level of detail to which the system is described, called coarse graining, is also necessary if it involves the loss of the information required for understanding the nature of the increase of entropy in irreversible processes. As a matter of fact, this finding is not surprising, because if we observe the filming of an event that takes place at the molecular scale, it is impossible to determine the direction of the macroscopic transformation of the overall gas.

To deepen this point, we can observe that, even in the context of classical physics, the chances are subjective, because they reflect our degree of ignorance about the details of the processes that we are observing. For instance, the introduction of the concept of probability in the assessment of the flipping of a coin reflects our inability to perform an accurate calculation by means of Newtonian mechanics of the trajectory of the coin until its fall, and that for lack of information on the initial data of the cast and on the environmental conditions. Knowledge of these is a privilege of the Laplace devil.

The situation, however, is completely different in quantum mechanics, since the probability of events comes from a genuine uncertainty about the behaviour of the world, and therefore its use cannot be removed by acquiring more information on the underlying phenomena. This awareness allows us to prove a more intriguing aspect of thermodynamics, because it expresses, through the second law, a declaration of impotence in our ability to transform all of the available heat in mechanical work. Quantum mechanics, which includes the uncertainty principle according to which the position and velocity of a particle cannot be simultaneously measured with accuracy, confirms the presence of an inescapable uncertainty inherent to microscopic phenomena. It has been shown that its violation would imply a violation of the second law of thermodynamics. Beating the uncertainty limit requires the extraction of extra information about the system, which requires doing more work on what is allowed by its state of disorder.

Let us now take a look at microbiology. Most of the involved objects have dimensions that lie at the boundaries of the scale at which quantum mechanics dominates, raising some disturbing aspects due to uncommon events, such as entanglement and tunneling. Why exclude the fact that they could play a significant role in the interpretation of some phenomena that are at the basis of life? Unfortunately, we can't neglect Richard Feynman, the legendary theoretical physicist, who said that nobody understands quantum mechanics. To address the problems involving nanoscale objects, it is therefore necessary to accept its mysteries and manage its paradoxes.

Finally, it must be remembered that the Boltzmann equation has proved to have a wide effectiveness in the description of the evolution of gases, pure and in mixture, even those with relatively complex molecules. For that, it is currently applied to the solution of technical problems such as those concerning the flow of fluids, or the diffusion of gas mixtures in various situations that are of interest in engineering, thus confirming that the hypothesis of molecular chaos is a good approximation, even if it is unable to catch the elusive nature of irreversibility. This is the curious fate of a theory that is successful in regard to that which concerns comparisons with experimental data, but is also a source of vibrant discussions for its theoretical foundations, in contrast to other theories that are accepted with enthusiasm, even if their results in regard to experience remain uncertain.

1.3   The Devil Is a Pawl

The energy of a gas of N molecules in equilibrium with the surrounding environment fluctuates around a mean value E with deviations ΔE described by a Gaussian function. It follows that

$$\frac{\Delta E}{E} = \frac{1}{\sqrt{1.5N}} \approx N^{-1/2}. \tag{1.1}$$

Therefore, the size of the energy fluctuations are:

– Comparable with the mean energy of an individual particle if N is equal to one,
– Negligible in macroscopic samples with N on the order of the Avogadro number.

Because, in small systems, the thermal fluctuations can lead to observably large deviations from their average behavior, it follows that they are not well described by classical macroscopic thermodynamics. Accordingly, we wonder whether it is possible to obtain mechanical work by exploiting the fluctuations of energy in a gas. One way to address this problem is found in the first of the volumes collecting the lectures of Richard Feynman, who, although famous for his contributions to quantum electrodynamics, did not hesitate to turn his interest towards peculiar issues. For instance, the one concerning the behaviour of a device, now called "Feynman's ratchet", borrowed from a machine invented by the Polish scientist Marian Smoluchowski in the wake of the Maxwell devil. The device, illustrated in Fig. 1.1a, consists of a cylindrical axis set in rotation by the effect of molecular collisions on blades grafted onto the axis. The rotation can take place in only one direction thanks to the presence of a pawl, which acts on a toothed wheel mounted in turn on the axis, thus transferring part of the energy randomly distributed in the molecular movements to the mechanical energy required for the shaft's rotation. In

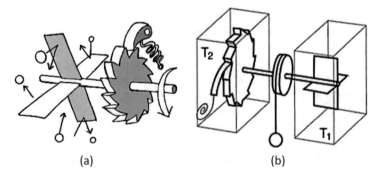

(a)   (b)

Fig. 1.1 (a) A cylindrical axis is set in rotation by the effect of molecular collisions on blades grafted onto the axis. The rotation can take place in only one direction thanks to the presence of a pawl, which acts on a toothed wheel mounted, in turn, on the axis. (b) The machine devised by Feynman in which two vanes at temperatures T_1 and T_2 are introduced to separate the part of the shaft on which the blades are grafted from the one containing the toothed wheel with the pawl

essence, the pawl would exercise the Maxwell devil's functions. If this could work, it would not violate the first law of thermodynamics, because the gas can be maintained at constant temperature by a continuous feed of thermal energy, but it would, however, violate the second law, because it would be drawing mechanical energy from a gas of constant temperature.

Feynman then devised the machine in Fig. 1.1b, more complex than the previous one, in which two vanes at temperatures T_1 and T_2 are introduced so as to separate the part of the shaft upon which the blades are grafted from the one containing the toothed wheel with the pawl. Then, he demonstrated that if the entire device is at the same temperature, that is, $T_1 = T_2$, the axis does not rotate continuously in one direction, but will rather move randomly back and forth. Therefore, it does not produce any useful work, because the pawl, being at the same temperature as the overall system, will undergo random motion, "bouncing" up and down.

With different temperatures, and assuming an exponential distribution $exp\,(-\varepsilon/k_BT)$ of the energies ε of the fluctuations, in agreement with the Boltzmann equation (Box 1.1), when the gas is at equilibrium, work is obtained. As a final result, the following expression of the efficiency η, i.e., the relationship between the work and the heat caught from the gas, is obtained:

$$\eta = \frac{work}{heat} = \frac{T_1 - T_2}{T_1}. \tag{1.2}$$

This finding is consistent with the analysis developed by Carnot in his book, which has subsequently been confirmed by an in-depth analysis performed at the microscopic scale. The preceding equation gives the higher value of the efficiency of a thermal machine that cannot be overcome. In real situations, as shown by Tolman and Fine, a term must be subtracted from the right hand side that accounts for the energy dissipation due to the creation of entropy as a consequence of the collisions between molecules, in accordance with the Boltzmann analysis summarized in Box 1.1, and at higher scale than that of the interactions between turbulent fluctuations.

In fact, the previous equation implies that the involved transformations take place through a sequence of equilibrium states. Despite being an oxymoron, the approach can be applied to the performance of approximate thermodynamic calculations if the awareness of its limit is accounted for. In fact, the previous expression provides the higher efficiency that can be obtained from a heat engine. Therefore, the following sentences written by Sir Arthur Eddington can be supported by belief:

> The law that entropy always increases, holds, I think, the supreme position among the laws of Nature. If someone points out to you that your pet theory of the universe is in disagreement with Maxwell's equations—then so much the worse for Maxwell's equations. If it is found to be contradicted by observation—well, these experimentalists do bungle things sometimes. But if your theory is found to be against the second law of thermodynamics, I can give you no hope; there is nothing for it but to collapse in deepest humiliation.

To conclude, it is worthwhile to recall that in his seminal talk "*There's Plenty of Room at the Bottom*", delivered in 1959, Richard Feynman offered a $1000 reward to the first person able to construct a motor that would fit into a cube 0.04 cm a side, not counting the wires and power source. This reward was claimed within a year by an engineer, William McClellan, who delivered to him a hard matter model of the micro-engine. But when turned off, the motor did what could be expected: nothing. In fact, the motor must be connected to a power supply to get any motion at all. But what about tinier engines of molecular size?

1.4   Swimming in Molasses

In 1828, the Scot Robert Brown published a book in which he summarized the observations conducted as to the behavior of particles present in the pollen of plants. By profession, he was a botanist who traveled extensively, collecting pollen grain of different kinds and with diameters of a few micrometers that, suspended in water and observed with a microscope, were subject to continuous disordered movements. At first glance, he thought he might be observing the "elementary molecules of organic bodies" expressing the life force itself. In these motions, which were later baptized as Brownian, it seemed that each particle had equal chances to move in any direction and that its previous state had no bearing on the future. These results cast a new light on the problem of the structure of matter that, throughout the nineteenth century, was the center of great controversy, the most visible manifestation of which was Boltzmann's suicide. Desperately looking for a theory? Of course, but it was formulated only 80 years later, when, in 1905, Albert Einstein provided an elegant explanation of how small molecules of water can impart movements to particles big enough to be observed under a microscope. In other words, the thermal energy associated with the motions of the molecules present in the liquid is transferred through the fluctuations to the motion of the particles, in apparent disagreement with the second law of thermodynamics. It is interesting to remember that Einstein, when developing this research, was not aware of the discovery of Robert Brown, because his specific purpose was to highlight facts that might confirm the existence of atoms. Then, his discussion stressed the fact that a suspended particle can be struck in various directions by the combined effect of small bumps from the surrounding smaller particles. All of that by taking into account that, as it was evidenced by Dean Astunian, with particles of nanometric sizes in water the viscous forces predominate, and therefore the motion is similar to that of a swimmer bathing in molasses.

Inspired by Brownian motion, in the first half of the last century, some mathematicians began to consider the description of phenomena in which the need to give a precise meaning to the presence of random movements was imperative. The main proponent was Norbert Wiener, Professor at MIT and founder of a transdisciplinary method for exploring the regulatory system known as cybernetics, with potential fallout in various fields, including biology, as will be illustrated later. The general approach, called stochastic from the Greek word for "haphazard", is a mathematical construct that models the real world through processes consisting of states that occur randomly. Their peculiarities are the absence of memory and the continuity. Something new under the sun? Not at all, because the stochastic methods were introduced in the nineteenth century and one mathematician who engaged in the subject was the afore-mentioned Pierre-Simon de Laplace, the repository of scientific determinism. His interest probably emerged when he realized the impossibility of giving life to his forward-looking demon.

Stochastic calculus includes an analysis of many natural processes involving random fluctuations, whose description is supposed to model independent and identically distributed shocks with zero mean. The standard approach of the stochastic evolution was borrowed from an equation that had been formulated by the physicist Paul Langevin in 1908 for the description of Brownian motion. A particle that is jiggling about because it is bombarded on all sides by irregularly moving water molecules should, of course, change its position with time. Because the collisions are random, each step is not necessarily related to the previous one, and thus the path is irregular, like the staggering walk of a drunken man coming out of a wine bar. Nevertheless, we can ask how far it is likely to get from the starting point after a given length of time.

A possible framework for the formulation of a mathematical model capable of describing the motion of an individual particle of mass m relies on the deterministic Newton equation, in which the product of the mass of the particle times its acceleration is put as equal to the sum of the applied forces, which are, respectively, the gravity and the viscous drag, due to some "effective forces" arising from its interaction with the other particles. This could be a job for the Laplacian devil, but, unfortunately, with the likelihood of obtaining a poor result, because the role of the random forces due to thermal noise cannot be neglected. For a particle with instantaneous velocity v, the equation describing its motion must be written as follows:

$$m\frac{dv}{dt} = F_{ext} - \mu v + F_{Br}(t), \tag{1.3}$$

where F_{ext} is the external force acting on the particle, μv is the viscous drag that is assumed proportional to the velocity of the particle, and finally, $F_{br}(t)$ is the random force that could be controlled by a Maxwellian devil. Except he does not exist.

One way to obtain a reliable solution is to take into account that the fluctuating forces come from occasional impacts of the Brownian particle with the molecules of the surrounding medium, so that the mean value of the involved energy can be put as equal to zero. Moreover, it can be assumed that the interaction force involved in the impacts varies in an extremely rapid way over time, so that it can be assumed that two successive collisions are not correlated. Following this approach, the average value of the square of the path x of a particle subject to Brownian motion can be calculated by applying the above equation.

$$\langle x^2 \rangle = \frac{2k_BT}{\mu}\left[t - \frac{1}{\mu}(1 - e^{-\mu t})\right]. \tag{1.4}$$

As an alternative to the previous approach, credited to Langevin, an equation due to Adriaan Fokker and Max Plank is also employed by making use of a distribution function of the particles subject to stochastic motions as if they were supermolecules. From the formal point of view, it is similar to Boltzmann's equation, except that the term that reflects the molecular collisions is replaced by another one that accounts for the correlation between fluctuations.

1.5 Down in the Bioworld

Motility is the hallmark of life, ranging from intracellular transport of macromolecules of about 10^{-9} m (10 nm) in size to the flight of birds. Movement being one of life's central attributes, any motile element is compelled to generate the forces needed for its displacement by converting some other form of energy into mechanical energy. Among the wide spectrum of possibilities, let us focus attention on what happens inside cells, which offer an impressive collection of molecular traffics. Also, the cell is the best starting point to undertake an investigation into microbiological systems, if attention is focused on the way in which special chemicals come together to form the complex dynamic structures that can be recognized as life.

Cells are divided into two kinds, called prokaryotic and eukaryotic, respectively. The former are bacteria and have independent life; they fulfill different, complementary physical and chemical functions, thus behaving as biological machines. The latter are involved, as aggregates, in the formation of the tissues of living organisms. Eukaryotic cells are more complex, thanks to the presence of an additional structure, as their DNA is contained in a membrane-bound nucleus. Each of them resembles a city immersed in water, with several compartments that have different functions, as illustrated in Fig. 1.2. Briefly illustrated, the nucleus is the repository of genetic information, particularly that which drives the protein synthesis that occurs in membrane-bound organelles called ribosomes, which behave as complex molecular machines. The obtained proteins are the main constituents of living organisms, by virtue of their structural and functional properties. The mitochondria, contrastingly, are tiny organelles that behave as powerhouses. There are almost 400 in every cell and they generate almost all of the energy employed in the cell's activities. Finally, the Golgi apparatus is responsible for modifying and packaging proteins for delivery to targeted destinations, similar to the post offices in our society.

As for machines produced by human technology, cells require constant renewal of their constituents. This operation implies the selective transport of the needed components from one side of the cell to the other. This occurrence generates traffic that takes place through a network of connections similar to those present in a modern metropolis. Inside the cellular liquid, called cytoplasm, a rich network of cables and tubes is present that supports the structure of the cell. Moreover, it also creates the connections between its different sectors by sustaining the carriage of substances, nourishment, and organelles, according to a logistics that is the subject of challenging investigations. The transportations are performed by means of ad hoc

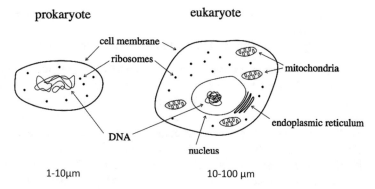

Fig. 1.2 A schematic description of prokaryote and eukaryote cells. Each of them resembles a city immersed in water, with several compartments having different functions, as illustrated in the figures. For details, see the text

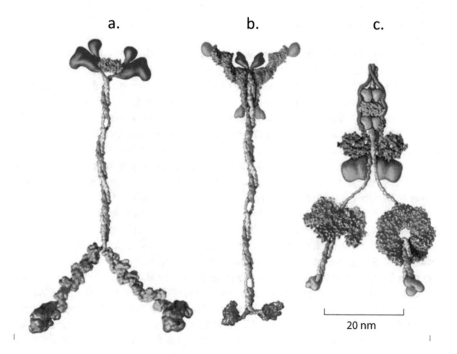

Fig. 1.3 Three typical molecular motors: (**a**) myosin, (**b**) kinesin, (**c**) cytoplasmic dyiein. The motor domains or heads are at the bottom. (*Vale RD. 2003. The molecular motor toolbox for intracellular transport. Cell 112:467–80*)

complex proteinic molecules, known as molecular motors, which are called, respectively (Fig. 1.3):

(a) **kinesin,** molecular motor that moves on microtubules, responsible for moving cargo.

(b) **myosin V**, molecular motor that transports cargo; myosin II is responsible for muscle contraction.

(c) **cytoplasmic dynein**, molecular motor that moves on microtubules.

Because of their nanoscale dimensions, on the order of 100 nm, they are subject to fluctuations of energy, which are so strong that they generate disordered Brownian motions. Nevertheless, they follow ordered paths along planned directions. This finding has been obtained through sophisticated and delicate experiences by measuring the displacement and the forces generated by the afore-mentioned protein motors. Specifically, a kinesin has been attached to a small silica bead that has been captured in the trap of a beam of photons coming from a laser. The photons together create a pressure from which a force capable of keeping a small particle from escaping the center of laser focus is created. These optical tweezers were used as a probe for measuring the kinesin movement on the microtubule attached to a glass surface.

What are the molecular mechanisms whereby such order might arise out of chaos? This is an intriguing question, because it reflects the main feature of living systems.

According to John Haldane (1892–1964), the distinguished British scientist known for his innovative contributions to evolutionary biology, life is a mixture of chemical processes, self-regulated thanks to the intervention of molecules that have the ability to perform special functions. The best candidates are, of course, proteins, it being unlikely that there are molecules with better properties anywhere in the entire Universe. Proteins are polymers of aminoacids, molecules with two functions, acid for the presence of a carboxylic group, and basic for the presence of an ammonium group (Fig. 1.4a). The chain of the proteinic polymer is illustrated in Fig. 1.4b, where the alternation between carbon and nitrogen atoms is highlighted. In living organisms, a wide range of proteins is present, which can be distinguished by the nature of the group R that replaces the hydrogen present in the formula of the aminoacid reported in Fig. 1.4b. The proteins carry out various functions, but attention will now be focused on those able to give a structure to the molecular motors, including their movements. As a typical example, we will look at the kinesin flowing along a microtubule that behaves as a rail. As it appears in Fig. 1.3, it is a small vertical bundle of proteinic molecules weaved together, with the two "heads" adhering to the bottom of the track, while two flexible filaments at the top connect it to a vesicle that contains the dragged load. But from where does the kinesin draw the energy required to move along the track and perform the work required to drive the load?

The cells are home to unceasing transformations involving energy fluxes, so that energy and life go hand in hand. The currency, or fuel, is the adenosine triphosphate, or ATP, a molecule (Fig. 1.5) having three phosphate groups linked end to end in a chain that, through reaction with water, produces one molecule of adenosine diphosphate, or ADP, plus one of phosphoric acid, P_i, and releases the energy that is employed in cellular activities as follows:

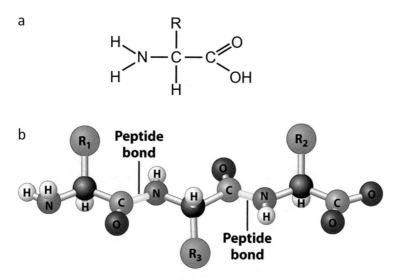

Fig. 1.4 (**a**) Chemical formula of a typical amino acid molecule. (**b**) Chain of a proteinic polymer in which the alternation between carbon and nitrogen atoms is illustrated. In a cell, the proteins carry out various functions, but here, attention is focused on its action as a rail of a molecular motor

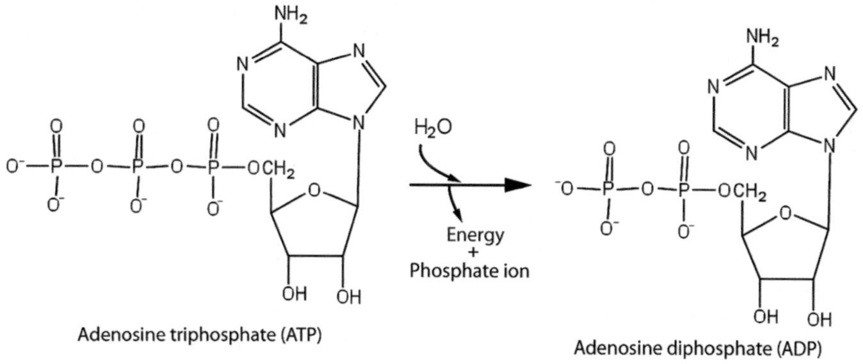

Fig. 1.5 Molecular structure of adenosine triphosphate (ATP) and adenosine diphosphate (ADP). The arrow indicates the hydrolysis reaction of ATP with water to give ADP plus phosphoric acid by releasing free energy

$$ATP + H_2O \rightarrow ADP + P_i + energy \quad (8.01 \text{ kcal/mole}).$$

Splitting away from the terminal phosphate group, the molecule releases a large amount of energy that is used to power most of the biological work, including that performed by kinesin and the other molecular motors. It is supplied in chemical form, because it can be involved in the breaking and forming of chemical bonds. The kinesin heads have pockets that house some ATPs, which, when set free, react with water through the previous reaction by releasing the energy that allows its

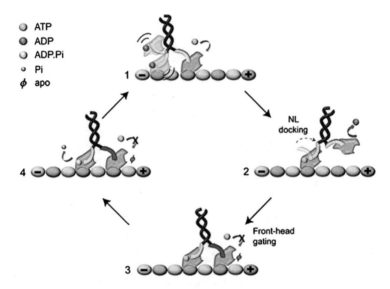

Fig. 1.6 Steps involved in the detachment and re-adhesion of kinesin on a microtubule. (1) The rear head, ADP-bound, weakly interacts with the microtubule. (2) ATP, binding to the front head, pulls its head forward. (3) The unbound head releases ADP and rebinds the microtubule ahead of its partner head. (4) The front head, red, is oriented backward, so that an ATP binding to this head is inhibited until the rear head hydrolyzes the ATP and releases it from the microtubule. (*Merve Yusra et al., Cell Reports 10, 1967–1973, March 31, 2015 a2015*)

movements. As shown in Fig. 1.6, it is able to facilitate the detachment of an end from the surface and its re-adhesion at a different location. Actually, all of the forces associated with a chemical act are short-range and nearly disappear at distances smaller than one nanometer, while one step of displacement in a motor protein usually extends to a few nanometers. Therefore, different principles are operative in molecular motors and follow their detachment from a microtubule. In fact, the breaking of a bond increases the kinesin's freedom, so that they become subject to the energy fluctuations of the surrounding fluid. The interaction of the kinesin with the microtubule can be expressed through a potential that has a periodic behavior along the path, according to the variation of the local composition of the microtubule surface. Therefore, the movement of the kinesin involves overcoming a succession of potential energy barriers, which, in a first approximation, can be equated to the teeth of a saw, while presenting an asymmetric behaviour, as is schematically illustrated in Fig. 1.7. A kinesin bounded with both heads behaves as if it were trapped in an hole of the potential energy, while if it is freed by the rupture of the bond, it acquires freedom and can move with equal probability in both directions as a consequence of the random bumps due to chaotic movements coming from the surrounding molecules. To describe its behavior, it is convenient to assume that it is controlled by a switch managed from the potential energy, as it appears in the afore-mentioned figure. When the pinning potential is turned off, a particle trapped at "0" begins to diffuse symmetrically because of thermal noise. Then, it begins to execute

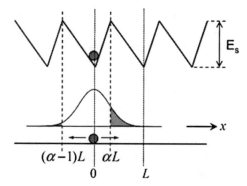

Fig. 1.7 Simple model of a Brownian ratchet through which a particle is subjected to a directed diffusion process on a surface. This is a consequence of the asymmetry of the potential, because the particle is more trapped at +L, to the right, than at −L. The loose motion of the coupled particle takes place thanks to the breakdown of an ATP molecule with the release of an ADP molecule. The particle, detached from the track, can freely diffuse, so that when it is reattached, displacement occurs. The process does not violate the second law of thermodynamics, because energy is supplied to the ratchet through the breakdown of the ATP molecule. (*Ping Xie, Int J Biol Sci 2010; 6(7):665-674. doi:10.7150/ijbs.6.665*)

a Brownian motion. When the potential is turned on, the particle is again trapped in one of the wells. Because of the asymmetry of the potential, it is more likely, at short times, that the particle is trapped at +L to the right than at −L, allowing for a net-directed motion. Thus, the mechanical movement is due to the energy of the chaotic motion of the liquid coupled with a chemical reaction occurring between the motor and the microtubule. The role of chemistry is to facilitate the choice of the fluctuations in one of the two directions, providing the energy required to drive the diffusional motion of the macromolecule without the intervention of the active force produced by the hydrolysis reaction. For this reason, the saw potential behaves as a "Feynman ratchet". In conclusion, Brown was right in assuming that the movements he observed were the 'life force', because most of the behaviour upon which life depends is driven by stochastic molecular motions. The preceding analysis has been focused on kinesin, but can be extended to the different moving motors present in living systems, as illustrated by Vologoskii ad Kolomeisky. The energy stored in chemical bonds is employed to generate directed forces for an amazing variety of tasks, but operates on the same principle: to trap Brownian fluctuations. In such a frame, two extreme types of motion can be distinguished: the former with the motor weakly coupled to the reference path, while in the latter, it is mainly subject to the effect of random collisions that drag it into the molecular storm. The theoretical analysis of their behaviour, pursued with the afore-mentioned approaches based on the Langmuir or Fokker-Planck equations, gives support to the previous description.

But what about the depletion of ATP molecules? They are rigenerated by means of the energy stored in particular molecules, such as the carbohydrates $(C_6H_{12}O_6)$ produced through photosynthesis, which catches the energy of the electromagnetic radiations generously sent to us by the sun. The process take place through a complex set of chemical reactions occurring in small molecular machines called mitochondria, where the carbohydrate molecules are transformed into carbon dioxide and water through the release of the energy employed to restore the ATP. Their role and their relevance will be illustrated in detail in Chap. 4. The term metabolism, derived from the Greek μεταβολή (mutation), indicates a set of chemical transformations, such as the previous ones, including the concomitant energy effects and their associated physical phenomena, that occur in the cells to ensure the preservation and renewal of the living matter. The overall cell metabolism occurs through a complex network of chemical reactions catalyzed by enzymes controlled by genes. It can be compared to a chemical engine that supports the cellular functions by converting raw materials into energy and into the molecules required to build biological structures.

Box 1.1 The Boltzmann Equation

A simplified form of the celebrated Boltzmann equation appears as follows:

$$\frac{\partial f(v,t)}{\partial t} = \int \left(f_2' f_1' - f_2 f_1 \right)(v_1 - v_2)\sigma(\omega)dv_2,$$

where $f(v,t)$ is the distribution function that specifies the probability that at time t, a molecule has a velocity v, 1 and 2 are the indexes of given molecules, and f' is the probability function after a collision.

The left hand side expresses the variation of the distribution function with respect to time, while the right hand side expresses the influence of molecular collisions.

$o(\omega)$ is the scattering cross-section, which, in turn, depends on the relative scattering angle ω and can be evaluated through a detailed analysis of the collision between the two molecules, a job that fits the Laplace devil.

At equilibrium, $f_2' f_1' = f_1 f_2$, and thus the right-hand side of the equation is equal to zero. Under such a condition, the probability that a molecule has a value equal to ε is proportional to $exp(-\varepsilon/k_B T)$.

References

Tolman C. Richard, Paul C. Fine. *On the Irreversible Production of Entropy*, Rev. Mod.Phys. 20, 1, 1948, 51-77.

Richard Feynman, Robert Leighton, Matthew Sands. *The Feynman lectures on Physics,* Addison-Wesley, Reading, 1963.

Gollub Jerry, David Pine. *Microscopic irreversibility and chaos*, Physics Today August 2006.

Astumian R. Dean. *Design principles for Brownian molecular machines: how to swim in molasses and walk in a hurricane*, Phys. Chem. Chem. Phys., 2007, 9, 5067–5083 l 5067.

Vologodskii Alexander. *Energy transformation in biological molecular motors,* Physics of Life Reviews 3 (2006) 119–132.

Ganhui Lan, Sean X. Sun. *Mechanochemical models of processive molecular motors*, Molecular Physics, Vol. 110, Nos. 9–10, 10–20 May 2012, 1017–1034.

Von Ballmoos Christoph, Alexander Wiedenmann, and Peter Dimroth. *Essentials for ATP Synthesis by F_1F_0 ATP Synthases,* Annu. Rev. Biochem. 2009. 78:649–72.

Kolomeisky Anatoly B., Michael E. Fisher. *Molecular Motors, A Theorist's Perspective*, Annu. Rev. Phys. Chem. 2007. 58:675–95.

Chapter 2
Entropy and Information

2.1 Back to Thermodynamics

The greatest American scientist of the period between the late nineteenth and early twentieth century was Willard Gibbs (1839–1903). Born in New Haven, Connecticut, he belonged to an old Yankee family that, since the seventeenth century, had produced distinguished clergymen and academics. After earning a Ph.D. at Yale, he traveled through Europe, attending lectures at different qualified Universities. Back home, in 1871, he was appointed chair of Mathematical Physics at Yale, the first such appointment in the United States. His research interest was focused on the extension of thermodynamics to chemical systems in which different phases (gas, liquid, and solid) are present. From such an effort, he wrote a monograph entitled "On the Equilibrium of Heterogeneous Substances", which begins with a quotation from Rudolf Clausius: *"The energy of the world is constant. The entropy of the world tends towards maximum."* Hereafter, Gibbs, rigorously and ingeniously, applied thermodynamics to the interpretation of certain chemical phenomena previously considered a collection of isolated facts and observations. The approach started from the definition of system, chosen as a portion of the Universe, whose behaviour can be characterized by a set of variables. Within this framework, Gibbs became concerned with "their private lives", which he addressed by setting forth the criteria of "their equilibrium and stability".

In thermodynamic equilibrium, there are no flows of energy and matter within a system and no composition change occurs. In Newtonian mechanics, energy is introduced as the property of moving masses, but in the nineteenth century, it became a unifying principle in the construction of the new sciences, including electromagnetism, thermodynamics and quantitative chemistry. In the following, it is advisable to express the energy of a thermodynamic system by means of a function H, called enthalpy, which is the sum of the kinetic and the interaction energies of the particles present in it, plus the product PV of the pressure, P, and volume, V, of the system itself. So, in the isobaric processes, the change of enthalpy is equal to the heat

© Springer Nature Switzerland AG 2018
S. Carrà, *Stepping Stones to Synthetic Biology*, The Frontiers Collection,
https://doi.org/10.1007/978-3-319-95459-2_2

transferred plus the work done in a change of volume at constant pressure. In fact, a force applied to an object determines its displacement, performing a mechanical work, W, given by the rule "force times distance". We already encountered, in the preceding chapter, the intriguing examples of the mechanical work performed by nanomotors on tubulines or by myosin in muscular movements. Hereafter, we will introduce other kinds of work present in living systems.

Let us start with a reversible process, which proceeds through a succession of equilibrium or near-equilibrium states, so that if it is reversed, it yields the starting material as it was before the transformation. The change of the thermal content of the system is determined by the fast motions of the microscopic atomic variables, which leads to an increase in the internal energy of the system without creating macroscopic work. Its variation, according to Clausius, is expressed by the product of the absolute temperature times the increase of entropy and, not being available to make work W, is called useless energy. Of course, it must be subtracted from the work that could be performed by the system as a result of a change of its total energy, expressed by the enthalpy. What remains is called free energy, because it can be used by the system to perform work through the slow variation of the macroscopic variables. The free energy G (from Gibbs, of course, as it was later baptized), is then defined as follows:

$$\text{available or free energy } (G) = \text{total energy } (H) - \text{unavailable or useless energy } (TS).$$

At the molecular level, the useless energy mainly contributes to the kinetic energy of the molecules and to their vibrational and rotational motions. The free energy can instead affect the behaviour of the atomic and molecular electrons through the breaking up and formation of intramolecular bonds, as well as increasing the number of available electronic states. The formation of new bonds can promote molecular organization by facilitating the formation of different types of molecule that contribute to the diversification of the system. For instance, a molecule can transform itself into some other isomers, having the same atoms, thanks to the exchange of free energy that allows for the formation of new structures with different contents of energy and entropy.

As has been argued by Rob Phillips and S.R. Quake in an article devoted to *The biological Frontier of Physics*, published in Physics Today, in 2006, at the characteristic distance at which molecular machines operate, the typical values of the different energies converge. It follows that, at the nanoscale, a particular situation is present concerning the conversion of the different forms of energy, which can be elastic, mechanical, electrostatic, chemical, and thermal. From their comparison, it appears that at characteristic lengths comprised of between 10^{-10} and 10^{-9} m, they have roughly the same values. This finding provides evidence for the possibility that molecules, and particularly biomolecules, can spontaneously convert their energies into several possible forms. In order to evaluate the exchanges of free energy involved in molecular transformations, Gibbs introduced the concept of "chemical potential", which was considered the Northwest passage, because it provided the link between thermodynamics and chemistry, as illustrated in Box 2.1. Its extension to biological macromolecules, in principle, is not different from that of any type of molecule, but the presence of higher complexity increases the underlying difficulties, because it exhibits a hierarchy of self-assembling structures ranging in size from proteins to membranes and cells. In a simplified approach, we can compare the cells to systems open to the exchange of energy, matter and information with the external environment. Inside, different events are occurring, including macromolecular movements, as with kinesin, chemical transformations, as in a metabolic network of chemical reactions, diffusional processes, and others. The energy source is Adenosine Tri-phosphate (ATP), which, as mentioned, is the fuel for most biological transformations, because it is the vehicle of free energy that, in living organisms, supplies the work for different activities. The transfer of free energy occurs through the familiar reaction of hydrolysis of ATP mentioned in Chap. 1. In living organisms, ATP is the vehicle used to supply the work, thanks to the magic effectiveness of free energy in changing the characteristics of a system by exploiting the available useful work in its various expressions:

Electric: transporting electric charges in tissues (ions) and nerves (electrons).
Chemical: synthesis and biosynthesis of chemical compounds.
Osmotic: active transport of molecules through membranes.
Superficial: increasing the surface of a system.

All of the listed forms of work are present in cells, which are very effective in taking advantage of their potential and interconvertibility. It is quite important then to deepen the mechanisms by which such energy release participates in the processes that take place at the microscopic level, such as the transfers of matter between fixed sites or the production of particular molecules through networks of chemical reactions. In the previous chapter, the transport of macromolecules within cells was discussed, while other processes will be considered with the intention of highlighting the way in which their occurrence contributes to favoring the emergence of order out of chaos.

2.2   The Statistical Character of Entropy

The statistical approach to thermodynamics is one of the theoretical pillars in the interpretation of the physicochemical world, including its extensions to biological systems. It was born in the second half of the eighteen century out of the efforts of Maxwell and, particularly, of Boltzmann, through the deepening of the insight into the relationship between the entropy of a system and the probability of its state. At present, it is commonly used in the description of the thermodynamic behaviour of large systems, by taking advantage of the knowledge of the atomic and molecular structures, and it is also applied, with some care, outside equilibrium by dealing with the description of irreversible processes such as flows of fluids and heat transfer driven by imbalances present in the system.

In 1901, Gibbs published a book of extraordinary compactness and elegance entitled "*Elementary Principles in Statistical Mechanics, Developed with Special Reference to the Rational Foundation of Thermodynamics*", in which he offered a new, powerful approach to the subject, introduced in a generalized and ingenious fashion. The approach was applied by means of classical mechanics, and the basic framework still stands, despite the modifications introduced by quantum mechanics, because the energy of a system can only assume a series of discrete values ε_i, each corresponding to one of its states. Let us consider a system in contact with a thermal reservoir at temperature T, which, at equilibrium, is equal to that of the system.

In this framework, the entropy is assumed to be a property of a collection, or ensemble, called canonical, of systems, each possessing the same volume, chemical composition and temperature as the system itself. If n_r is the number of ensemble members in the r-th state with energy ε_r, then their distribution among all possible states is given by $f_r = n_r/\Sigma n_r$. Then, it can be shown that, consistent with a constant value of internal energy, entropy assumes a maximum equilibrium value if

$$S = -k_B \sum_r f_r \ln f_r \quad \text{being} \quad f = \frac{e^{-\varepsilon_r/k_B T}}{\sum_r e^{-\varepsilon_r/k_B T}} = \frac{e^{-\varepsilon_r/k_B T}}{Z}. \tag{2.1}$$

Z, is called the partition function and depends on the temperature and the structure of the molecules, as well as the parameters that characterize their vibrational and rotational motions. k_B is called the Boltzmann constant, and its value is given in the Table of the physical constants. Moreover, any property Y of the system can be evaluated as the mean value of the statistical ensemble:

$$< Y >= \sum_r Y_r f_r. \tag{2.2}$$

The preceding equations are at the summit of statistical thermodynamics. Climbing down, the behaviour of matter can be evaluated starting from molecular properties, whereas at the top, the concept of thermal equilibrium can be deepened. For the simplest system, constituted by an ensemble of non-interacting point mass molecules, Ω is the number of their possible distributions among the system states. Because all have the same probability, it follows that $f_r = f = 1/\Omega$. Then,

$$S = k_B \ln \Omega. \tag{2.3}$$

In other words, the more states that are available, the higher the entropy will be. The previous equation was proposed by Max Planck, following the insight of Boltzmann. For this reason, it is engraved on Boltzmann's tombstone in the central cemetery in Vienna.

2.3 Out of Equilibrium

Is there a basis for any teleology in the second law of thermodynamics? Not really; any "designed" feature in the Universe must increase entropy, as stated by Clausius. The only reliable teleology is the forecast by Lord Kelvin concerning thermal death. Luckily, that is faraway! But what about the possibility that the features of natural transformations, at least locally, could trigger order and organization? At the limit, according to somebody, life itself? Thermodynamics is the science of change, but most of the situations considered up to now refer to reversible transformations occurring through a succession of artificially forced equilibrium states. Actually, the occurrence of a transformation implies the presence of a finite difference in temperature, pressure or chemical potential, which are the driving forces. At the final state of a transformation, in a system with constant internal energy, their values must be uniform everywhere, consistent with the maximum value of the entropy. Irrespective of the end point, the interest is towards the characteristics of the transformations, specifically in regard to their rates and eventually the rate of production of entropy. These problems were faced in 1930 in the USA by the Norwegian-born scientist Lars Onsager (Nobel prize in physics), by focusing attention on the case of small deviations from equilibrium, so that simple linear mathematical relationships could be adopted for expressing the rates of the irreversible processes.

Let us consider a system subject to a transformation attributed to a couple of forces X_1 and X_2, due, respectively, to the presence of differences in the temperature and in the concentration of a chemical component, with respect to the equilibrium values. Their influences on the rate of the process occurring are respectively expressed as a function of the temperature gradient and of the concentrations of the chemical components through a physical quantity called chemical potential, whose formal definition is reported in Box 2.1. Two coupled processes with flows J_1 (energy over time) and J_2 (moles of component i over time) occur, whose rates are expressed as follows:

$$\begin{aligned} J_1 &= L_{11}X_1 + L_{12}X_2 \\ J_2 &= L_{21}X_1 + L_{22}X_2. \end{aligned} \tag{2.4}$$

The coefficients L_{11} and L_{22} are proportionality constants, linearly relating each specific flow to its force, while L_{12} and L_{21} are cross-coefficients whose values depend on the extent of the coupling between the two processes, thus reflecting the way in which the force of one process affects the flow of the other. Close to equilibrium, a principle called microscopic reversibility is introduced by establishing that

$$L_{12} = L_{21}. \tag{2.5}$$

Onsager postulated that this symmetry condition could be applied with confidence to situations that were not very far from equilibrium, as was experimentally checked for that which concerns the diffusional and heat transfer processes in diluted gases and condensed systems. The approach is generalized to systems of many coupled processes by means of a set of linear equations similar to the previous one:

$$J_j = \sum L_{jk} X_k,$$

                                                                                    (2.6)

where, for the cross-coefficients, the afore-mentioned symmetry rule is still applied. The extension of the approach to biology looks reasonable, because it captures a characteristic property of living systems of coupling energetically efficient processes. Nevertheless, the problem of how far from the local equilibrium it can be applied to living systems remains open. In 1950, the Nobel laureate Ilya Prigogine, Russian-born but living in Brussels, together with Paul Glandsdorff, attempted to extend the linear approach of Onsager to the more interesting, non-linear regime, by stressing the role of chemical reactions whose rates customarily are given by non-linear algebraic expressions with respect to the concentrations of the chemical components involved.

In order to illustrate the outcomes of the investigation, let us focus attention on a system open to the exchange of matter and heat with the environment that can be compared to a well-mixed chemical reactor that maintains a mixture of reacting compounds far from equilibrium. In a first approximation, it can be adopted to describe the behaviour of a biological cell. By assuming a constant temperature, the analysis is focused on the material transformations occurring inside the system. The model equations are formulated through the material balance of each component, by accounting for the contributions resulting from their exchange with the environment and their disappearance or formation as a consequence of the chemical reactions. For each of them, the following conservation equation must be applied:

Variation time = sum of the inlets and outputs
                          + variation due to chemical reactions.

So, for component I, it can be written as follows:

$$\frac{dC_i(t)}{dt} = \sum_{ext} \Phi_{ext} C_i^{(ext)} + R_i(T, C_j),$$

                                                                                    (2.7)

where the left-hand side expresses the rate of change of the concentration C_i of component i expressed in moles per unit volume, while the two terms at the right-hand side express the contributions to such a variation due, respectively, to the exchange with the environment associated with the fluid fluxes Φ_{ext} and the variation of component i per unit time and volume due to chemical reactions. In other words, the preceding equation expresses the way in which the derivative of the concentration of component i depends on the rates of the reactions in which it is involved, without being affected by diffusional processes thanks to the imposed stirring. R_i potentially depends on the concentrations of all of the components present, but practically, only a few are involved. On the whole, by applying Eq. (2.7) to each component, a system of ordinary differential equations (ODEs) is obtained whose integration yields the evolution in time of the different concentrations.

The most interesting cases are the ones in which non-linearities are present in the mathematical expressions of the rates R_i. In fact, linear systems have been the stronghold of science for about 300 years, while non-linear systems, related to complex situations, frequently with unexpected results, have been neglected, mostly because of the difficulties of solving the inherent equations. Particularly, if the non-linearities are due to the presence in chemical systems of autocatalytic processes, occurring when one product of a reaction increases the rate of transformation of its own products, a rich variety of patterns describing the evolution of the concentrations, as a function of space and time, emerges (see for instance the book of Grégoire Nicolis). Thus, the ingredients required for the creation of complexity turn out to be at our fingertips. Most natural phenomena evolve smoothly, rather than by abrupt changes, and therefore, it is reasonable to model them by neglecting their variation over time, at least for a limited period of time. Such a stationary situation is called 'steady state' if it is assumed that the actual values of temperature and concentrations will no longer change in the future. Accordingly, the derivatives on the left-hand side of the balance equations are put as equal to zero, so that a system of algebraic equations is obtained whose solution yields the concentrations of the reactants under steady conditions. That is, in a situation in which all of the involved variables, such as species concentrations, temperature and fluid flow, do not depend on time in spite of the ongoing processes, because, notwithstanding the presence of a fluid flow through the entire system, there is no accumulation of mass or energy. The afore-mentioned algebraic equations being non-linear, different solutions are obtained, each corresponding to different set of values of the reactants' concentrations. After having discarded the physically unreliable solutions, we have to make a choice between the residual solutions. The more obvious approach is to look for their instability, so that after a perturbation due to a fluctuation, they are subjected to a transition towards another state of the system, while stable solutions recover the original state instead. This problem has been approached by Prigogine and Glandorff on thermodynamical grounds, in an attempt to identify a quantity that has the same pivotal role of entropy for the equilibrium states. Following a heuristic approach suggested from the symmetry rule (2.5) of linear thermodynamics, they proposed that the stability corresponds to a minimum rate of entropy production. Equipped with this criterion, they explored the role of the external medium, which manifests itself through the values of a suitable control parameter λ, typically, the fluid flow rate or the temperature, whose value determines the state of the system.

The analysis of typical systems reveals the behaviour of a steady state versus λ illustrated in Fig. 2.1. At a critical point, with $\lambda = \lambda_c$, the state becomes unstable and changes sharply, because a bifurcation starts. In other words, the steady state splits into two branches corresponding to a choice between two possible states of the systems. This interesting outcome is enriched by the subsequent intervention of further bifurcations that offers an interpretation of the evolution towards an increasingly higher level of complexity. This occurrence, which develops at a significant

Fig. 2.1 A typical bifurcation diagram. As the control parameter λ is increased, a state loses its stability and two new branches emerge at $\lambda \geq \lambda_1$. These branches, in turn, lose their stability at a secondary bifurcation point $\lambda = \lambda_2$, and so on

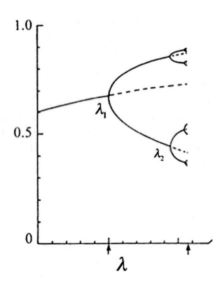

distance from equilibrium, corresponds to the formation of structures whose complexity has opened up many expectations in investigations into the self-organization of chemically reacting systems. In conclusion, the thermodynamic approach seemed to provide evidence that there would be no need for new natural laws in the description of the self-organization processes occurring in nature. Complex structures can appear far from equilibrium beyond the limit of stability.

Actually, a lot of work has been done on non-equilibrium steady states, so that, regrettably, in the second half of the previous century, some concerns about the relative stability of the states of a system far from equilibrium appeared. In particular, Rolf Landauer, a German-American physicist who made important contributions to different areas of the thermodynamics of information processing, has shown that the principle of minimum entropy production can only be expected to apply for steady states that are not far from equilibrium. Therefore, without specifying the exact domain of its validity, it cannot be considered a universal principle, and thus cannot be applied to state selection out of equilibrium. In this analysis, Landauer's work was reminiscent of the pioneering researches performed by the Russian scientist R.L. Stratonovich.

Moreover, other researchers have examined the steady state distribution reached after a long time, as well as their relaxation times, indicating the difficulties in stating their role in the emergence and evolution of biological systems. Skepticism then arose as to the employment of the word "self-organization" in connection with the appearance of life as a consequence of the tendency of biological forms to become more complex.

In conclusion, the fascinating problem of self-organization looked like a "hard nut to crack", because the relevant role that genetics and molecular biology can play in Darwinian evolution cannot be forgotten.

2.4   Shaping the World

How did the leopard get its spots? Why is the black-and-white pattern on the skin of a cow haphazard and asymmetrical? Still more challenging, how does the morphogenic phenomenon involved in the differentiation of cells blossom? Of course, in approaching these fascinating problems, the peculiar features of the rates of the chemical reactions must be accounted for, because they contain the spatial and time structures of the systems. Moreover, the diffusional processes must also be accounted for, because the simple well-mixed reactor is not suitable for facing such a problem.

The attempts to offer an interpretation of morphogenic processes, widely diffused in nature, was addressed by Alan Turing in 1952, in a pioneering paper entitled *"The chemical basis of morphogenesis"*. Turing has become an unsung hero, because his work on cryptanalysis during the Second World War is now recognized as the single most important contribution to the Allied Victory. His outstanding work on formalisation of the concepts of the algorithm is considered a milestone in mathematics and opened the route towards the development of computer science, with important fallout for biological science, as will be discussed later. In the aforementioned paper, which was his last, he faced the ambitious problem of explaining the shape of living embryos using knowledge about the kinetics of the underlying chemical processes. He was inspired by gastrulation, the process through which the developing cell goes through a series of segmentations, passing from a spherical shape to more complex geometric configurations. The instability arising from the rupture of the initially spherical symmetry is attributed to a chemical reaction. The corresponding model implies a competition between an activator that acts at a short distance and an inhibitor that acts at a long distance. The presence of substances called morphogens that enhance the activity of particular genes that repress others is invoked. The typical equation of the Turing model is reported and discussed in Box 2.2, while in Fig. 2.2 it is illustrated the calculated evolutions as a function of time, of the concentrations of the components, present in a system in which the competition between an activator and an inhibitor is present. Similar equations are called "chemical pattern generators" for their ability to produce a variety of patterns present in nature. Actually, even if the designs that emerge from Turing's equations are impressive, the scientific community of biologists has welcomed the issue with skepticism, because it seems to usurp the prerogative of genetics to control the morphogenetic processes. Morphogens were not discovered during Turing's lifetime, but rather, in 2012, a group of researchers at King's College London showed that Turing was right, by showing that two chemicals control the formation of ridge patterns inside of a mouse's mouth. The question is still the object of discussion.

In reality, as emphasized by Stephen Ornes, despite the controversies as to their meaning and interpretation, non-equilibrium systems surround us. This is typically due to the presence of chemical reactions, because their rates depend exponentially on temperature and, often, in a non-linear way, on the concentrations of the reactants involved. For instance, one intriguing experiment that displays unexpected behavior

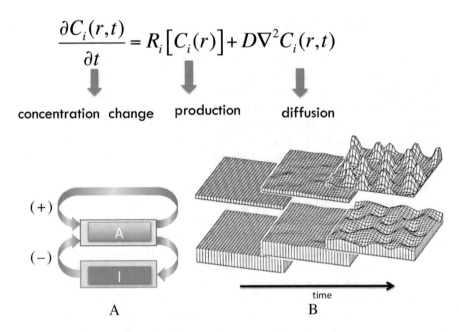

$$\frac{\partial C_i(r,t)}{\partial t} = R_i\big[C_i(r)\big] + D\nabla^2 C_i(r,t)$$

concentration change   production   diffusion

Fig. 2.2 The Turing reaction-diffusion model, which provides a mechanism for pattern formation in nature, when the interactions between an activator (A) and an inhibitor (I) are operating. The terms present in the equation correspond, respectively, to the processes associated with the variation of the concentration of component i as a function of time t and of space coordinates x

is the popular Belousov-Zhabotinskii (BZ) reaction, which, performed in a dish, exhibits red and blue oscillations propagating as spiral wave fronts. The changing colors are caused by alternating oxidation-reduction processes in which cerium changes its oxidation state from Ce(III), producing a magenta solution, to Ce(IV), producing a blue solution, or vice versa. Another significant example is fluid convection. If the bottom of a cup of water is heated, the heat dissipates through the liquid in stripe-like shapes. But if the bottom of the cup is continuously warmed, the heat dispersion gives rise to intricate patterns that are not easily predicted by means of the laws of physical chemistry. These amazing behaviours appear in more than just fluid convection, because driving a fluid far from equilibrium does not simply lead to turbulence, but also to complex structures. Actually, in the studies of non-equilibrium processes, the equations that describe how fluids move, widely known as the non-linear Navier–Stokes equations, are applied to investigate what happens far from equilibrium. Nevertheless, the underlying mechanics that drive pattern formation is not yet well understood. For this reason, high-precision, detailed experiments are in progress for the purpose of establishing a unifying theory of non-equilibrium thermodynamics that can provide a molecular explanation for the non-equilibrium behavior. Recently, cells have attracted interest and research, because they are hotbeds of non-equilibrium activities. In their cytoplasm, a cyto-skeleton is present, whose mechanical properties are dominated by the protein polymers that form microtubules and some intermediate filaments. Moreover, a network of coupled chemical reactions maintains a chemical non-equilibrium state, by feeding the corresponding mechanical framework with energy. In this contest, molecules self-organize into complex machineries and patterns so as to drive functions, such as the described intracellular transport and cell locomotion. Of course, to claim that the inside of a living cell is dynamic is a truism, but the highlighting of the mechanisms of the underlying motions requires the employment of techniques that enable investigation of the cell dynamics over a wide range of temporal and spatial scales.

In conclusion, without a doubt, non-equilibrium thermodynamics plays a significant role in the diffusion processes that occur in cytoplasm, in the transportation of material within cells with molecular motors and in cell metabolism. For all of those reasons, it is gaining concrete interest among the scientists who are chasing the secrets of life.

2.5 The Role of Information

Why don't you call it entropy? In the first place, a mathematical development very much like yours already exists in Boltzmann's statistical mechanics, and in the second place, no one understands entropy very well, so in any discussion, you will be in a position of advantage.

With these words, in 1948, John von Neumann, the legendary mathematician, answered a question put to him by Claude Shannon, a young IBM engineer, attempting to offer an approach to measuring the information contained in a message.

But what is information and, above all, what is the aspect that makes it so revolutionary? Our awareness, as evidenced by James Gleick, tells us that there is too much of it, because we are continuously bombarded with a huge flow of information, diffused by media that is more and more technologically advanced, but, unfortunately, less and less transparent and reliable.

Claude Shannon had the privilege of giving a quantitative surface to the concept of information, publishing, in 1940, an article that now occupies a prominent position in the Ghota of scientific literature, perhaps without immediately realizing that the concepts developed therein contained the seeds of a theory that, in a short time, would spread out until it captured a leading role in culture and modern technology. Its title, "*Mathematical Theory of communication*", indicates that his analysis was particularly addressed to the technology devoted to the transmission of messages, which, thanks to its marriage with a great invention that debuted in the same year, the transistor, would result in a dramatic development within the world of communications.

The content of a message can be expressed through a set of binary decisions, as in the game in which, by means of a succession of yes or no answers, it is possible to identify a specific card in a deck or a person in a group. The information is expressed by the number of binary decisions needed to choose between the two alternatives. Each of them is referred to as a bit, short for Binary Digit.

Let us consider, as a first example, a situation in which all of the choices have the same probability, by focusing on the case of eight objects, such as playing cards, out of which one in particular must be chosen. If the deck is split into two parts of four cards each, the first binary choice concerns the location of the one of the two that contains the reference card. Proceeding further to the division into two cards each, it comes out that the desired card is identified through three binary choices. By generalizing, if I choices of Ω objects must be made, it follows that $\Omega = 2^I$, and thus

$$I = \log_2 \Omega, \tag{2.8}$$

I being the information requested to identify the system.

The situation becomes more complicated if the events have different probabilities. Suppose, however, that they can be separated into different groups containing n_r events, each with a probability $p_r = (n_r./\Omega)$. In this case, the information is evaluated by means of the following expression:

$$I = -\sum_r p_r \ln p_r, \tag{2.9}$$

which is similar to Eq. (2.1), introduced to evaluate entropy in the framework of Boltzmann's and Gibbs' approaches to statistic thermodynamics. It is easy to acknowledge that if probabilities p_r have the same value, Eq. (2.8) is obtained.

The last equation, which is the highlight of Shannon's theory, clarifies Von Neumann's proposal aimed at usurping one of the most important outcomes of thermodynamics. Actually, its comparison with Eq. (2.3) indicates that information has the same roots as entropy, because it measures our ignorance as to the energy distribution among the different states of a thermodynamic system about which we know only the values of some macroscopic quantities, such as temperature, pressure and volume. Therefore, the uncertainty about the missing information regarding the occupation of the Ω microstates of a thermodynamic system with a given energy is a good metaphor for entropy.

2.6 The Physics of Neglecting

We already encountered Rolf Landauer in his contributions to the behaviour of the systems with two or more competing states of local stability, but now, it is imperative to mention that he was also a pioneer in the area of information handling. In fact, it is his famous principle, now applied to computers, that communication cannot avoid use of a small amount of energy. Thus, information cannot be an abstraction, because each bit is linked to a physical situation, such as the holes in a piece of perforated paper, the state of a neuron, the spin up or down of a particle, etc. In other words, a bit cannot exist without some embodiment. Moreover, the information is encoded, processed and transmitted by *physical means*. In fact, physical systems such as capacitors or spins are used for storage, sound waves or optical fibres for transmission, and the laws of classical mechanics, electrodynamics or quantum mechanics dictate the properties of the afore-mentioned devices. *"Information is inevitably Physics"* is the title of the last paper by Landauer. This awareness has significant implications for understanding what information is by indicating that it is not a purely mathematical concept, but that the properties of its basic units are dictated by the laws of physics. Then, if information is registered by physical systems, and all physical systems can register information, we can combine Eqs. (2.3) and (2.7) to obtain

$$I = \frac{S}{k_B \ln 2} = \frac{energy}{k_B T \ln 2}, \tag{2.10}$$

where S is the entropy of the system, while on the left-hand side, the relation between entropy and energy has been introduced. In other words, the amount of information, measured in bits, that can be registered by any physical system is bonded to their energy content.

The afore-mentioned researches of Landauer from 1961 started by taking note of the most interesting fact to come from Eq. (2.10), namely that the energy associated with the unit of information, that is, one bit and corresponding to $I = 1$, is equal to $k_B T \ln 2$. This is nothing more than the energy required by the Maxwell devil to control the movements of molecules. Further development relies on the familiar concepts of reversibility and irreversibility, by observing that the logical operations involved in the management of information do not imply the dissipation of energy. Indeed, when a bit flips from zero to one, or vice versa, the information is preserved, so that the process is reversible and entropy is unchanged, because no energy is dissipated. In agreement with the second law of thermodynamics, which implies the presence of irreversibility, Landauer formulated the hypothesis that computation is necessarily connected to a dissipation of energy whenever a bit is discharged. Then, the only irreversible processes involved in computation are those in which information is eliminated, so that cancellation achieves a leading role. In particular, he showed that the erasure of information causes the release of $k_B T \ln 2$ of heat per bit in the environment. Forgetting takes work. In fact, the Maxwell devil needs information to affect the motion of one molecule present in a gas, but in order to be ready for further operation, he must discharge it by producing an amount of entropy equal to $k_B \ln 2$. Charles Bennet, a young pupil of Landauer at IBM, analyzed every kind of computer, real or abstract, by confirming that a great deal of computation can be done with no energy cost. Dissipation occurs only when information is erased.

In conclusion, a sober answer to the initial question as to what information is leads to its embodiment in things like computers and electrons until the brain's neural firings. Edward Fredkin, a digital physicist, who collaborated with Richard Feynman and was involved in various technical businesses, investigated the twilight zone concerning the interface of computer science and physics. He went further than common expectancies by substantiating that information is more fundamental than matter and energy, because atoms, electrons, and quarks ultimately consist of bits of information, like those that are the currency of computation in a personal computer. Their behavior, and thus the behaviour of the entire Universe, is governed by a suitable programme. This idea, which, at the limit, suggests that we are living in a simulation or control system of some kind, has been the subject of science fiction novels and films, such as *The Matrix*. Nevertheless, if the boundary conditions are associated with solid concepts, such as those of energy and entropy, it is interesting to explore the possibility of producing knowledge about the capacity of the Universe to generate its pervasive complexity.

Box 2.1 Chemical Potential

Chemical potential is obtained by differentiating the energy with respect to any of the substantial components of the system:

$$\mu_i = \left(\frac{\partial G}{\partial n_i}\right)_{T,P,n_j},\tag{2.11}$$

n_i being the number of moles of the i-th component and G the Gibbs free energy. An immediate application is given by the evaluation of the change of free energy due to the variation in the physical conditions:

$$W = \Delta G = \left(\sum_i n\mu_i\right)_{final} - \left(\sum_i n_i\mu_i\right)_{initial}.\tag{2.12}$$

In it, the mole variation of each component in going from the initial to the final state is due either to mass transfer with the environment or to chemical reactions. A convenient approach that has been adopted is to express it as follows:

$$\mu_i = \mu_i^0(T,P) + RT \ln x_i + g_{int}(x_i).\tag{2.13}$$

On the right-hand side, the first term is the molar free energy of the pure component i at temperature T and pressure P, respectively. The second term is the contribution resulting from the mixing process by assuming an ideal mixture in which the molecules of each species, having molar fractions x_i, are compared to massive points without reciprocal interactions and randomly distributed in the space. The final term reflects the influence of molecular sizes and their reciprocal interactions in the description of the behavior of real systems, as illustrated in the book of Lewis and Randall revised by Pitzer. It

(continued)

Box 2.1 (continued)
is the more subtle and engaging term to obtain, because its evaluation implies the application of the many-body calculation technique that requires an accurate knowledge of intermolecular forces.

Box 2.2 Bifurcations and Stability of Steady States
The dynamic behaviour of a network of chemical or enzymatic reactions, occurring in biological systems, can be conveniently analyzed and described by means of a set of differential equations, each one expressing the material balance of each component. The values of the concentrations identify the state of the system, while their variation over time characterizes its evolution. If the reacting mixture is well mixed, the values of the concentrations are the same at any point of the whole system. Then, its behavior over time is formalized by a set of ordinary differential equations (ODE), such as (2.3). Each can be written, concisely, as follows:

$$\frac{dC_i(t)}{dt} = F(C_1, C_2, \ldots, \lambda), \tag{2.14}$$

where λ indicates the external, or control, parameters, such as the reagent flow rate or the temperature, which can be modified by the external world. The model could be considered deterministic in the sense that the same initial conditions produce the same results in repeated simulations. Actually, the previous statement must be accepted with caution, because in non-linear systems, small differences in the initial conditions could also produce significantly divergent paths.

Under stationary conditions, the terms on the left-hand side of the previous equations disappear, and therefore a system of algebraic equations appears, usually non-linear, whose solution provides the values of the concentrations corresponding to the various possible steady states of the system. The results depend, of course, on the values of the control parameters λ, whose variation affects the numbers of the steady states and their characteristics. Typically, it corresponds to the insurgence of bifurcation processes by increasing the value of λ, as indicated in Fig. 2.1. At small values of λ, only one solution is present, and so the corresponding steady state can be considered an extrapolation of the equilibrium conditions. A common property present in the afore-mentioned steady states is the stability, called asymptotic, that is essentially determined by the response of the system to perturbations. A basic result of the stability theory establishes that the afore-mentioned stabilities are equal to those of the linear part of Eq. (2.14), which can be written as follows:

(continued)

Box 2.2 (continued)

$$\frac{dC_i(t)}{dt} = \sum_j L_{ij}(\lambda)C_j(t), \qquad (2.15)$$

$L_{ij}(\lambda)$ being the linear coefficients of the series development of Eq. (2.14). It follows that the problem can be addressed with the known methods of calculus. The approach, obviously valid only for examining the behaviour of the system in proximity of the steady state, is employed to test its local behavior after a perturbation.

The general solution to the linearized system of differential equations is expressed through a combination of exponential terms:

$$C_i = A_1 e^{\sigma_1 t} + A_2 e^{\sigma_2 t} + \ldots \qquad (2.16)$$

The values of coefficients $A_1, A_2 \ldots$ depend on the initial conditions, while the exponents σ, named after Liapunov, depend on the values of the external parameters. It follows that if all of them are negative, the terms on the right-hand sides of the previous equation approach zero, and so the system, after the perturbation, recovers its initial state. Therefore, the state is stable. If only one of the exponents is positive, the corresponding term in the previous equation diverges. Then, the system is unstable, and a transition towards another state can occur.

When spatial gradients of the concentrations of the cellular components are present, their transport due to the diffusional processes must be accounted for. In this more realistic situation, a system of partial differential equations (PDE) must be employed, in order to describe the evolution in time and space of the component concentrations. This approach has been used to analyze the pattern formation in the development of biological organisms, expressed by the values of the concentrations at the different points and as a function of time, as illustrated in Fig. 2.2. Typically, the evolution of a reacting system is described by a diffusion-reaction model expressing the material balance of each component by means of the following equation:

$$\frac{\partial C_i(r,t)}{\partial t} = \sum_{ext} \Phi_{ext} C_i^{ext} + \nabla^2 C_i(r,t) + R_i[C_i(r)], \qquad (2.17)$$

$C_i(\mathbf{r}, t)$ being the concentration of component i ay a point defined by the vector r(x,y,z) at time t. D_i is its diffusion coefficient, which depends on the structures and the molecular interactions of the involved molecules. The Laplacian of the concentration of the component under examination, $\nabla^2 C_i$,

(continued)

Box 2.2 (continued)
present on the right-hand side, expresses the driving force of the diffusive process, expressed by the rate at which the diffusion rate at a point differs from the average value at the points surrounding it.

In conclusion, differential equations can describe nearly all systems undergoing change, so that their presence is ubiquitous in science and engineering, as well biology. Very often due to the complexity of the investigated systems, an analytical solution of the obtained equations cannot be pursued, and therefore numerical methods are customarily employed. Increasingly complex systems of differential equations can, at the present, be solved with suitable programs written to run on available computers, including common PCs. Different programs are available for approaching the problems, and a convenient choice concerns the use of MATLAB programs, designed for scientific computing by making use of a variety of numerical operations, including the solution of the non-linear algebraic equation systems capable of investigating the presence and characteristics of stationary states.

References

Shannon Claude. *A Mathematical Theory of Communication*, The Bell System Technical Journal, Vol. 27, pp. 379–423, 623–656, July, October, 1948.

Nicolis Grégoire. *Introduction to Non linear Sciences*, Cambridge, 1995.

Gleick James. *The Information*, FOURTH ESTATE, London, 2010.

Brown Julian. *Minds, Machines and Multiverse, The quest for Quantum Computers*, Simon and Shuster, New York, 2002.

Leff Harvey S., Andrew F. Rex. *Maxwell's Demon, Entropy, Information, Computing*, Adam Hilger, Bristol,1990.

Machta J. *Entropy, information, and computation*, 1077 Am. J. Phys., Vol. 67, No. 12, December 1999.

Lewis Gilbert Newton, Merle Randall, Revised by Kenneth Pitzer and Leo Brenner. *Thermodynamics*, Mc. Graw-Hill Book Company, New York, 1961.

Nelson Philip. *Biological Physics*, Freeman, New York, 2008.

Landauer Rolf. *Inadequacy of entropy derivatives in characterizing the steady state*, Phys. Rev. A, 18, 8, 1975, 636-638.

Ornes Stephen,. *How nonequilibrium thermodynamics speaks to the mystery of life*, PNAS I January 17, 2017 I vol. 114 I no. 3 I 423–424.

Chapter 3
A Computer Called the Universe

3.1 A Controversial Origin

On a shining morning in the summer of 1948, Ralph Alpher, a 27-year-old graduate student, was walking in his most elegant suit towards the main building of the Princeton University to discuss his Ph.D. dissertation. The event had generated a lot of public interest because of the unusual topic, prominently mentioned in the local newspaper: he was going to debate aspects of the world's creation.

Not everyone found it appropriate to analyze the creation of the world through a scientific approach, it being an intrusion into a domain previously reserved for God. For this reason, Alpher received letters in which many people claimed that they would pray for his soul and his redemption. What is following is a fashinating story, as it is told by Kragh Helge.

The topic of the research was suggested by George Gamow, a creative scientist, Russian-born but an emigre to the USA, where he became a professor at George Washington University, who believed in a cosmic scenario characterized by a decrease in temperature associated with the expansion of the Universe. The idea was inspired by the development of the theory of general relativity credited to the Russian physicist Alexander Friedman, who, in 1922, published the exact solution of the cosmological equations introduced by Einstein. The subject was revamped a few years later by Georges Lemaitre, a Belgian priest and physicist, who came to the theory with a stunning perspective, because, for the first time, the idea that there was a day of creation emerged into the scientific mainstream. It was without experimental data, but in 1929, Edwin Hubble published a paper in which he reported his observation that the galaxies are rushing away from us.

Actually, Gamow's interest was also triggered by a publication from 1927 by the Norwegian geochemist Victor Goldschmidt, in which the results of an extensive survey of terrestrian, meteoric and astrochemical spectroscopic measurements, shown in Fig. 3.1, were reported by enriching the content of the Mendeleevian table of the elements. The relevance of these results confirmed the generalized

© Springer Nature Switzerland AG 2018
S. Carrà, *Stepping Stones to Synthetic Biology*, The Frontiers Collection,
https://doi.org/10.1007/978-3-319-95459-2_3

Fig. 3.1 Distribution of the elements in the Universe, published in 1937 by Victor Goldschmidt, obtained by an extensive survey based on terrestrial, meteoritic and astrochemical measurements. Goldschmidt realized that this data reflected the cosmic history of creation

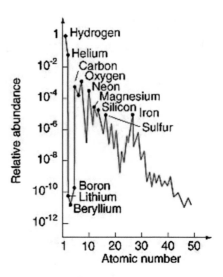

opinion of the scientific community that Mendeleev should be regarded as a foregrounding figure in the history of chemistry, with a position similar to that of Linnaeus in biology, who proposed the criteria of classification of living organisms by opening the way for the formulation of the Darwinian evolutionary hypothesis. A similar role was emerging from the afore-mentioned geochemical results.

Gamow was eager to examine how chemical elements could be formed by supposing the presence of a succession of nuclear reactions, starting from an initial neutronic state of the world called "Ylem", a term invoked to baptize a hypothetical original plasma. Going through a set of neutron captures by increasing the atomic mass of nuclei, and beta decays through the emission of an electron, by increasing their atomic numbers, it is potentially possible to obtain all of the elements present in the periodic table of Mendeleev. Then, the rate of increase of the mass number of a nucleus is equal to the difference of the rates of its build-up and the rate of further transformation, and consequently, the time evolution of the different elements is described through a set of consecutive reactions. From the solution of the differential equations that govern their balances, after a suitable adjustment of the value of the initial neutrons' density, it was possible to interpolate, with reasonable approximation, the distribution of the different elements present in the Universe.

The results were published in a paper in the Physical Review having as its authors Ralph Alpher, Hans Bethe and George Gamow resulting in its being called "The alphabet paper", while its content became known as the $\alpha\beta\gamma$ theory. Actually, Bethe did not contribute to the research; the presence of his name was the result of Gamow's sense of humor.

However the $\alpha\beta\gamma$ theory met with serious difficulties, because, as pointed out by Enrico Fermi and Anthony Turkevich, no stable nuclei exist with atomic masses 5 and 8. To overcome this difficulty, Edwin Salpeter proposed the following mechanism, called (3-α) involving three nuclei of helium:

$$^4\text{He}+^4\text{He}\rightarrow^8\text{Be}$$
$$^8\text{Be}+^4\text{He}\rightarrow^{12}\text{C} + \Upsilon.$$

Unfortunately, this process was revealed to be inefficient, because of the ^8Be instability. Such an embarrassing situation was overcome thanks to the penetrative intuition of Fred Hoyle, a creative and outstanding English scientist, if also a controversial character, who realized that, in the presence of an excited state of the carbon atom, the second step could have been sped up by many orders of magnitude. The existence of such a resonant state has been further experimentally confirmed by William Fowler, a nuclear physicist, director of the Kellogg Radiation Laboratory at Caltech.

Nevertheless, the new path devised to build heavy elements required high temperatures so that it was necessary to find furnaces in the Universe that were sufficiently hot and dense to forge the elements. In other words, it seemed a good moment to take a fresh look at the stellar interior. This message was grasped by a team that included Hoyle himself and Fowler, plus Margaret and Geoffrey Burbidge (B^2HF), who developed a theory on the formation of elements into stars, in their various phases of development. According to them, in a main star sequence, hydrogen burs to helium, which produces further nuclei, so that, following nuclear chemistry, the elements' distribution takes place as illustrated in Fig. 3.2. More recent researches on the composition of the stars are consistent with the afore-mentioned nucleosynthetic processes. Hydrogen and helium are still most abundant, but 79 other stable elements also exist in appreciable abundance.

At very high temperatures, the stars collapse through an explosion that scatters their materials into space in the form of gas and dust. Further condensation gives rise to new stars and planets. In conclusion, Gamow and his wife Barbara summarized the situation through the following new version of Genesis:

New Genesis

In the beginning, God created radiation and ylem, without shape, into which nucleons were rushing.

And then, let there be mass 1,2,3 ... up to 92 elements.

But when he looked back, he had missed calling for mass five, and, naturally, no heavier elements could be formed.

And then he said, let there be Hoyle, and he told him to make heavy elements in any way he pleased.

And Hoyle decided to make heavy elements in stars and to spread them around by supernova explosion.

Barbara and George Gamow

Paradoxically, Hoyle strongly disapproved of any theory involving the appearance of the Universe from nothing, "*like a party girl jumping out of a birthday cake*". (Ironically, he was responsible for coining the term "Big-Bang" during a BBC radio program in 1949.) Instead, he postulated the existence of a "Creation Field", with a negative pressure consistent with the energy conservation and expansion of the Universe. Therefore, at that time, two cosmological models were competing:

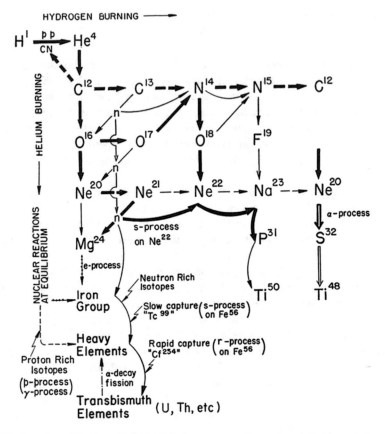

Fig. 3.2 Overview of the formation of stellar elements according to the original formulation of the B^2HF theory. (Burbidge et al. 1957)

– The steady state, in which the apparent expansion occurs in a homogeneous and isotropic way
– The standard Big Bang, in which, after an explosion, a sharp expansion occurred, followed by a gentle expansion still in progress.

In 1964, Arno Penzias and Robert Wilson, two researchers at Bell Labs, were experimenting in Holmdel, New Jersey, with a supersensitive horn antenna originally built to detect echoes bounced off of balloon satellites. To measure these faint radio waves, they had to eliminate all recognizable interference from their receiver, such as the effects of radar, radio broadcasting and the heat in the receiver itself. Despite this, they heard some astonishing radio signals, so that they wondered if they had made a mistake. Was the signal radio noise from nearby New York City? Or still the result of a defect in their instrument? After further careful analysis as to anything that could cause the excess thermal noise, the suspicion emerged that, after a year of experiments, they were detecting cosmic background radiation, an echo of the universe at a very early moment after its birth.

The Big Bang theory was officially born.

Nevertheless, though today, it is generally accepted that the Universe started with a big bang, the standard view in regard to that which concerns the formation of the elements is owed to the incorrect steady state theory of Fred Hoyle, who was killed by the accidental discovery of the cosmic background radiation.

3.2   The Atomic Universe

The Universe was born 13.7988 (±0.037) billion years ago. Its origin is unknown, but any speculation cannot discard the question raised by Gottfried Wilhelm von Lebnitz, *"Why is there something rather than nothing?"*, which was defined by William James as the *"darkest question in all of philosophy"*. *"To be or not to be"* wrote Shakespeare, while Saint Augustine, in a still more disturbing thought, argued that God created hell for those who investigate the deep mysteries of nature. Therefore, for us, it is convenient to approach the subject with caution.

Everything started in a radioactive era, with a huge amount of energy in the form of electromagnetic radiations, which turned into particles as the temperature decreased, as a consequence of a subsequent quick expansion. This happened through a series of transitions, similar to the passages from gas to liquid or to solid, common in familiar matter. All of this happened over a very short period of time, and the description of the processes involved is pursued by means of the "standard model" of elementary particles, that is, a mathematical framework used to describe all the known particles and forces, except gravity.

The standard model includes a set of rules that allow for the calculation of the probabilities of the possible transitions, consistent with the physical conditions. Particularly as the hot temperature decreased, our familiar matter was created, made out of electrons gravitating around nuclei, containing protons and neutrons, in turn performed by quarks and gluons that bind the quarks together, similarly to the Russian dolls known as *matryoshka*.

The whole affair is summarized by a single equation, known as Lagrangian, that, just to get an idea, is reported as follows in the cups of tea that you can buy at CERN.

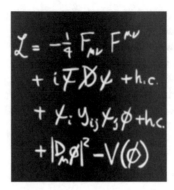

Starting from the Big Bang, as the Universe cooled, quarks and gluons got together to form nucleons, protons and neutrons, followed by the atomic nuclei, each with a mass determined by the number N of nucleons and a charge determined by the number Z of the protons. The explosion associated with an extremely quick expansion was followed by a gentle expansion, which is still underway, associated with a temperature decrease. In the first three minutes, 98% of all existing matter was produced in an amount sufficient to trigger the nuclear reactions required to produce the elements of the periodic system, as described in the previous paragraph. The neutralization of the positive charge of nuclei from the gravitating electrons produced atoms, or positive ions if one or more electrons were lacking. We are thus authorized to borrow the following sentences from the first volume of Feynman's lectures on Physics:

> If, in some cataclysm, all scientific knowledge were to be destroyed, and only one sentence passed on to the next generations of creatures, what statement would contain the most information in the fewest words? I believe that all things are made of atoms—little particles that move around in perpetual motion, attracting each other when they are a short distance apart, but repelling upon being squeezed together. In that one sentence, you will see that there is an enormous amount of information about the world, if just a little imagination and thinking are applied.

This is an excellent enlightenment for any further development that will proceed, going back to Mendeleevian classification. In the periodic table, before 1937, four of the 92 elements were missing, having atomic number Z equal to 43, 61, 85, 87, respectively, while the last practically stable element is bismuth, Bi, with atomic number 83. The six subsequent elements (from 84 to 89), which are radioactive and present in nature in insignificant amounts, are followed by a set of radioactive elements: thorium, protactinium and uranium (90, 91, 92).

An urgent question emerges: *"Is it possible that the ingenuity of man could replace God in the creation of new elements?"* In other words, could the periodic table be considered an unfinished work, like the Tower of Babel? The synthesis of the element 43 by irradiating the molybdenum (42) with deuterium gave us the possibility of filling the spaces of the other missing elements and to extend the periodic table beyond uranium.

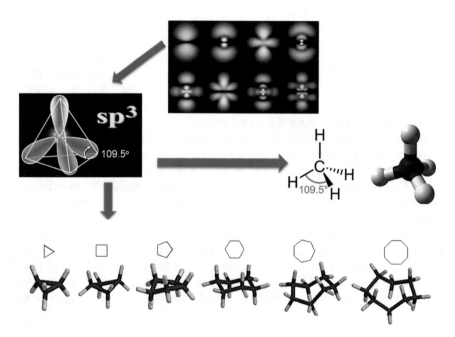

Fig. 3.3 Starting from the three-dimensional shapes of the hydrogen atom orbitals, the sp³ hybrid orbitals are built. When carbon combines with four other atoms, the basic tetrahedral shape presents in the molecules of organic chemistry and bio-organic chemistry

But what is the limit of the periodic table? The stability of nuclei has been studied by comparing them to liquid drops and testing their sensitivity to fission due to the competition between the surface forces coming from the attractive nuclear forces between protons and neutrons and electrostatic repulsions between the charged protons. Atomic nuclei with more than roughly 100 protons are unstable, no matter how many stabilising neutrons they contain, but theoretical researches have revealed the presence of an "island of stability", in which heavy enough nuclei might hold together. That island, though elusive, has been explored, and the synthesis of the element with $Z = 114$ has confirmed the hypothesis as to the existence of such a small nuclear stability area in a sea of nuclei of very short life. The element baptized Flerovium (Fl) was obtained by bombarding a target of Plutonium (Pu) with accelerated nuclei of Calcium (Ca), an achievement that was pursued through a dizzy scaling down from the Universe to a laboratory (Flerov Laboratory for nuclear research, Dubna), having man as its protagonist.

Therefore, for a chemist, the world is made of nuclei and electrons, whose motion in agreement with quantum mechanics is described by a wave function Ψ whose square gives the probability of finding the electron at a given point. Because the electrostatic attraction of the protons confines the motion of the electrons to a particular region of space, a series of patterns of stationary waves emerges, as shown in Fig. 3.3. Such forms define the structure of the molecules, because the formation of bonds between atoms occurs in the directions in which the overlap between the lobes of the wave functions is higher, as illustrated in the figure itself.

The most important ingredient for atomic and molecular stability is the Pauli exclusion principle, which avoids the collapse resulting from the Coulombian attraction between electrons and nuclei. In fact, each electron is confined in a "private volume" due to the presence of an "energy of exclusion" that generates a repulsive effect. For molecular objects having nanometric dimensions, the repulsion is balanced by the electrostatic attraction exerted on the electrons from the nuclei. It can be shown that the size of a molecule with n electrons is approximately given by $0.6n^{1/3} \times 10^{-8}$ cm. The application of such a relation reveals that atoms and molecules have relatively large structures, full of open space, with well-defined central nuclei. This is the main reason for the existence of chemistry: but what is its relevance in the Universe?

The survey on the different types of object present in the Universe offers a picture in which their sizes at different scales are determined by the competition between attractive and repulsive forces, but quite interesting is the presence of a "magic box" that contains the range of structures exhibiting properties characteristic of organized and complex systems whose investigation is of relevance. All belonging to living systems.

3.3   The Molecular Universe

Matter in the Universe is present in galaxies that are gravitational systems. Besides stars, a significant amount of cool matter, at about 10 K, is dispersed into wide clouds, in which both gas, at a density between 10^2 and 10^9 (atoms/cm^3), and small dust particles (0.1 μm radius) are present. A high-energy radiation including protons and atomic nuclei is also present, still of mysterious origin, despite the fact that some results indicate that a significant fraction of such primary cosmic rays originate from supernovae explosions.

To be exhaustive, so called dark matter must also be mentioned, whose existence and properties are inferred from its gravitational effects, such as the motions of visible matter. Because, up to now, it has not been revealed by means of the commonly used methods of physics and chemistry, we can neglect its possible influence on chemical reactions. In addition, several molecules are present in diverse astronomic environments, ranging from nearby objects in our solar system to distant sources in the early universe. Molecules are associated with dense and cool neutral interstellar matter in our and other galaxies, both in gaseous phase and in the solid phase of the tiny dust particles. Wherever molecules are found, they are probes of the conditions of their environments and of the lifetimes of the sources. But most importantly, non-terrestrial molecules are of interest for what they tell us about the build-up of molecular complexity throughout the universe.

Spectroscopy, that is, the use of the absorption, emission, or scattering of electromagnetic radiations by atoms or molecules and their ions, is an important tool for the investigation of chemical reactions, as well as in many other areas of science and technology. It is, of course, an indispensable instrument for astrophysical science, because, from the information obtained on the transitions from a lower level to a higher level with the transfer of energy from the radiation field, or vice versa, we can get information on the nature of molecules present in space and of their movements and transformations. Usually, transitions within the molecular rotational energy levels belong to the far infrared and microwave spectral region, transitions within the molecular vibrational energy levels belong to the infrared spectral region, and finally, the transitions between the electronic energy levels belong to the visible, or ultraviolet, spectral region.

The wealth of molecular spectra yields detailed physico-chemical information capable of characterizing the nature of the multitude of molecules present in the cosmic environment, and moreover, the high-resolution of the spectra associated with the rotations and vibrations of molecules tells us about the density and temperature of the gas (see the cited paper of H. Eric and E.F. van Dishoeck). The simulations through models, including the probable networks of the chemical transformations, provide interesting and useful information on the molecular evolution towards increasingly complex structures. And lastly, because the calculated molecular abundances depend on time, as well as on physical conditions, the obtained results can yield information about the histories of their sources.

The large number of molecular species that has been detected via high-resolution spectral features range in size from 2 to 13 atoms, mainly carbon and those belonging to organic chemistry. Most of them are common terrestrial species (H_2O, NH_3, alcohols,....), while others give rise to a witches brew of positive ions (H_3^+, HCO^+, H_2O^+) and radicals (C_3H, C_3H_2,...) that are very reactive species for the presence of unpaired electrons.

Of the over 150 different molecular species detected in the interstellar and circumstellar media, approximately 50 contain 6 or more atoms. As mentioned, most of them are organic in nature, so that carbon atoms dominate the cosmic chemistry. Such organic molecules are mostly unsaturated radicals of the form CnH, (n = 2–8) and nitrogen containing compounds of the general formula HCnN; n = 3, 5, 7, 9, 11.

The presence of the polycyclic aromatic hydrocarbons (PAHs), composed of multiple benzene rings, is relevant. The simplest is naphthalene, having two aromatic rings, while the three-ring compounds are anthracene and phenanthrene. The PAHs, such as those shown below, and related species, such as their dehydrogenated, ionized and substituted derivatives, are omnipresent in the interstellar medium.

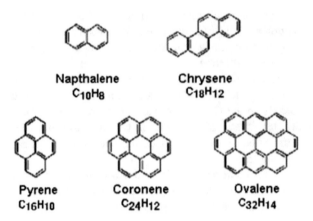

It has been suggested that PAH-like species account for up to 30% of galactic interstellar carbon and have been implicated in the astrobiological evolution of the interstellar medium, providing the nucleation sites for the formation of carbonaceous dust particles.

3.4   The Birth of Chemistry

The amazing abundance of chemical compounds in the inhospitable regions of the Universe requires a deepening of their origin by identifying the possible and reasonable mechanisms through which they are formed. The triggers for the formation of such a complex and diversified multitude of chemicals are the ionization reactions of the most abundant species, H_2 and He, due to the incidence of cosmic radiation CR:

$$H_2 + CR \rightarrow H_2^+ + e^-$$
$$H_2 + H_2 \rightarrow H_3^+ + H.$$

The obtained trihydrogen cation, also known as protonated molecular hydrogen, is one of the most abundant ions present in the Universe. It is the simplest example of a three-center, two-electron bond system, and its stability in the interstellar medium is due to the low temperature and density. Its role in the gas-phase chemistry is unparalleled by any other molecular ion, because H_3^+ is a strong protonating agent that promotes different reactions, such as:

$$H_3^+ + CO \rightarrow HCO^+ + H_2$$
$$H_3^+ + N_2 \rightarrow HN_2^+ + H_2$$

As the gas-phase chemistry proceeds, atoms and molecules are physisorbed into dust particles through weak non-bonding van der Waals forces, or chemisorbed through valence forces. Then, surface reactions occur whose rates are proportional to the interactions between the adsorbed species, by accounting for the saturation of the active sites present on the surface. This model, called the Langmuir-Hinshelwood, is familiar to the people who work in heterogeneous catalysis. In this framework, the dominant water-ice species is assumed to be produced by the sequential hydrogenation of O atoms landing on a grain:

$$O \rightarrow OH \rightarrow H_2O,$$

and a similar mechanism is invoked for the conversion of N atoms into NH_3 and C atoms into CH_4. Also, carbon monoxide produced in the gas and adsorbed onto grains can be hydrogenated into alcohols, such as methanol, through a sequence of reactions involving atomic hydrogen:

$$CO \rightarrow HCO \rightarrow H_2CO \rightarrow H_2COH \rightarrow CH_3OH.$$

As mentioned, polycyclic aromatic hydrocarbons and related species play a key role in the astrochemical evolution of the interstellar medium, but the formation mechanism of even their simplest building block, that is, the benzene molecule, is still elusive.

It has recently been demonstrated, with crossed molecular beam experiments, combined with electronic structure and statistical calculations, that benzene can be synthesized via the reaction of the ethynyl radical C_2H and 1.3-butadiene, the simplest hydrocarbons with a couple of conjugated double bonds:

$$C_2H + H_2C = CH - CH = CH_2 \rightarrow C_6H_6 + H.$$

This reaction portrays the simplest representative of a reaction in which an aromatic molecule with a benzene core can be formed from acyclic precursors.

But how are molecules synthesized in space, since the existing low density and temperature are not associated with rapid chemistry? Indeed, in a gas phase, the rate of a chemical reaction depends on the frequency of molecular collisions, which, in the interstellar contest, is relatively low. But as a matter of fact, the rate of a reaction between a positive ion and a neutral species is enhanced by the electrical polarization effect through which a slight relative shift of positive and negative electric charge in opposite directions within a molecule takes place. This slight separation of charges makes one side of the molecule somewhat positive and the opposite side somewhat negative, improving its aptitude to lead a chemical reaction. In the case under examination, the rate becomes about two orders of magnitude higher and does not depend on temperature. As mentioned, the interstellar ice growing on the solid dust particles present in cold dense clouds can be the substrates for chemical transformations, including polymerizations. Their presence, which involves the occurrence of reactions at the very low temperatures of the cosmic dust, implies the intervention of a particular mechanism. V. Goldanskii, in 1976, showed that the polymerization of formaldehyde can occur through a tunneling process promoted by the movement of the monomer onto the end of the polymer chain. This effect is a consequence of the wave behaviour of molecular particles, which allows them to penetrate the potential energy barrier, hindering the contact between the molecular reacting centers. Because the tunnel effect is not affected by temperature, the polymerization reactions can also take place in a very cold environment. Of course, our main interest is devoted to the so-called prebiotic molecules, in so much as they can offer some enlightenment on the life processes. Some are those necessary for the maintenance of life, having water in a primary position, and the others are produced by living organisms or through biological processes such as methane. Higher relevance must be given to the detection of molecular species having structural elements in common with those found in living organisms, such as the:

- First interstellar "sugar", that is, glycolaldehyde CH_2 OHCHO,
- Acetamide, CH_3CONH_2, that is, the largest interstellar molecule with a peptide bond typical of aminoacids,
- Amino acetonitrile NH_2CH_2CN, that is, a direct precursor of glycine, the simplest amino acid.

These findings teach us that carbon, oxygen and other elements forged in the nuclear furnaces of stars can get together in interstellar space to form not only simple molecules such as water, carbon monoxide and ammonia, but also some of the complex molecules of the terrestrial chemistry where life has evolved. Work is in progress, and the perspectives are promising in regard to exciting new discoveries. But very astonishing is the fact that, despite its apparent absence of hospitality with respect to chemistry, the Universe looks like a crowded zoo of cosmic organic compounds.

Inevitably, a question arises: is there room for a scenario in which life came from space? Panspermia, from the Greek πᾶν (all) and σπέρμα (seed), is the hypothesis that life exists throughout the Universe, distributed by meteorites and, maybe, by spacecraft in the form of contamination by microorganisms. It is an old idea that has inspired many writers of science fiction to create fun and pathos through characters like ET, or terror with the images of the fierce invaders from Mars by Herbert G. Wells. The first known mention of the term is in the writings of the fifth century BC Greek philosopher Anaxagoras, but it began to assume a scientific veneer through their revamping by the Swedish chemist Svante Arrhenius and, more recently, by Fred Hoyle with Chandra Wickramasinghe, an Indian astrobiologist.

Despite the attractiveness of the idea, Panspermia does not actually address the question of how life started everywhere and, at least up to now, it does not provide evidential support for discharging our blù planet from the responsibility of having originated life. And for us, it is not an alibi for discarding the puzzles involved.

3.5 Back to Earth

In 1953, Stanley Miller, a graduate student of the Nobel laureate Harold Hurey at the University of Chicago, undertook a creative experiment reminiscent of the speculations that Darwin expressed in 1871 in a letter addressed to his friend Joseph Hooker:

> If (and oh what a big if) we could conceive in some warm little pond all sorts of ammonia and phosphoric salts, light, heat, electricity, present, that a protein compound was chemically formed, ready to undergo still more complex changes...

For the first time, the hypothesis was advanced that the various aspects of life could converge to a single point of origin. Actually, Darwin was reluctant to publish his views on life's origin, but nevertheless, he prefigured a concept that became, as time passed, increasingly popular, until it was incorporated into the mediatic culture and popularized as the dominant idea in any speculation on the origin of life.

Also, even though the word biochemistry was officially introduced in the second half of the eighteenth century, studies and researches concerning the chemical processes occurring in living organisms started earlier than that. Of relevance, of course, was the proposal of 1780 by Antoine Lavoiser that combustion is similar to respiration because both processes need oxygen. Its conceptual importance was that, for the first time, a vital process was interpreted without reference to living organisms, in contrast with the dominant concept of "vitalism", by which the present organic molecules can only be produced by other organisms.

In this contest, an important achievement was pursued in 1828, when the German chemist Friedrich Whoeler synthetized the urea by obtaining, for the first time, through chemical manipulations, a chemical compound produced in living organisms.

$$(NH_4)_2CO \rightarrow NH_2CONH_2.$$

In 1893, Eduard Buchner, a German chemist and physician, gave the first demonstration of the possibility of performing alcoholic fermentation in the presence of a cell-free extract of yeast cells, thus demonstrating that the presence of living cells was not needed for fermentation. Despite such achievements, the belief in the existence of a "vitalistic" force was still present at the beginning of the nineteenth century. In fact, such a philosophy was not only the territory of professional biologists, but eminent physicists also believed that the chemistry of atoms and molecules could not explain the core of living matter.

The idea that life emerged in a pond rich in chemical ingredients became popular only after 1920, thanks to the independent promotion of two scientists, the Russian Alekxander Oparin and the Brit J.B.S. Brit Haldane. The latter was a central figure in evolutionary biology and an excellent communicator of science. The famous science fiction novel *"Brave new world"* by Aldous Huxley, published in 1932, was mostly inspired by the publication in 1924 of the pamphlet *"Daedalus, or Science and the*

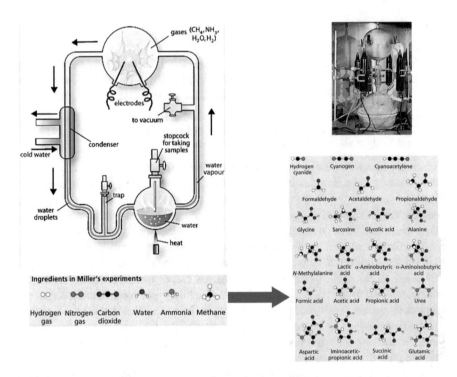

Fig. 3.4 The Stanley Miller experiment was performed with the equipment illustrated in the figure. Methane (CH_4), ammonia (NH_3) and hydrogen (H_2) were sealed in a sterile 5-L glass flask placed at the summit, and connected to a 0.5 L flask at the bottom, half-full of liquid water, which was evaporated by allowing the vapour to enter the larger flask. Continuous electrical sparks were fired between the electrodes to simulate an atmosphere of lightning

Future", in which Haldane spread a prophetic optimism on a near future when biology would become applied to eugenics so successfully that diseases would be eradicated.

A steep change in the subject was accomplished 30 years after, in the same year in which Crick and Watson published the structure of DNA, thanks to the afore-mentioned experiment by Miller in which, by means of the equipment illustrated in Fig. 3.4, a simulation of the chemical reactions occurring in conditions similar to the ones present on the young earth was attempted. Methane (CH_4), ammonia (NH_3), and hydrogen (H_2) were sealed in a sterile 5-L glass flask put at the summit, connected to a 0.5-L flask at the bottom, half-full of liquid water that was evaporated, allowing the vapour to enter the larger flask. Continuous electrical sparks were fired between the electrodes to simulate an atmosphere of lightning.

At the end of one week of operations, the process was stopped and the reaction mixture was analyzed by means of paper chromatography. The presence of five amino acids was identified in the solution, including glycine, α-alanine and β-alanine. After Miller's death in 2007, some sealed vials preserved from the original experiments were examined and the presence of more than 20 different amino acids produced in the original experiments was found, more than the 20 that naturally occur in life.

For his amazing finding, Miller became a celebrity, and the excited press did not lose the occasion to make all sorts of exaggerated comment. The results seemed to provide stunning evidence that life could arise out what Oparin and Haldane had called the "primordial soup". More imaginative people maintained that scientists, like Dr. Frankenstein, would shortly create living organisms in a laboratory. Unfortunately, or fortunately, this approach did not work. At least, it hasn't up to now.

Further evidence suggested that Earth's original atmosphere might have had a different composition from the gas mixture used in the Miller experiment. Moreover, in the following years, the initial recipe was modified, while the need for more sophisticated mechanisms for biomolecule emergence was emphasized.

Thus, despite the advancements, we are still at the initial problem: what is the molecular basis of life? Life is a remarkable network of molecules, catalysts and reactions present in a cell, but it is operating in an elusive way, because the molecules and the reactions are not alive. How can the complicated components of life, when put together, give rise from the disorganized chaos of the prebiotic earth to the network of reactions called a cell? In other words, what is the complexity level that must be reached in order to generate life?

Because its main peculiarity is the emergence of behaviours that could drive a crucial transition, the previous framework seems to imply:

- Non-linearity of the rate expressions,
- Strong gradients of reactant concentrations (C_i)
- Interconnectivity of the agents (I)
- Energy E flow through the system
- Cycling of energy flow

So, what might a mathematical recipe of emergence look like? Maybe an inequality, such as the following proposed by Robert Hazen:

$$complexity \leq f[C_i, \nabla C_t, I, \nabla E(t)], \tag{3.1}$$

by interpreting "*complexity*" as a critical numerical value to be overcome. But both the number on the left-hand side and the explicit form of the right-hand side expression are not known, and so the reason for the transition from the abio- to the bio-world remains an enigma. It is tempting to invoke the intervention of an evolutionary mechanism, based on the dialectic alternation between mutation and selection. However, at the molecular level, the thermal fluctuations are too weak to amplify the self-replicating features so as to yield perceivable stable patterns. Actually, the origin of biological matter implies the emergence, through a complex network of prebiotic chemical reactions, of structures with new properties, but also with an information content that can be duplicated. Finally, it can be observed that, despite its popularity, the concept that life arose from a soup containing the organic components present in living organisms has been subject to criticism. In fact, if we were to take a living organism and homogenize it so as to destroy its cellular structure, but leaving the molecules intact, no form of life would ever arise, because in that perfect organic soup, the carbon, nitrogen, oxygen, and hydrogen are at equilibrium. Life, in fact, implies the presence of continuous and energy-releasing chemical reactions in a system far-from-equilibrium, and, in particular, physico-chemical situations that deserve further deepening.

3.6   The Computing Universe

In a science fiction story from 1956, entitled "*The Last Question*", Isaac Asimov takes us through an intellectual adventure experienced by two young men involved in the construction of an increasingly larger computer aimed at exploring the behaviour of our galaxy, taking advantage of the increase in computer size, and consequently of its progressively greater amount of power. Moreover, the intriguing question: "*How can the net amount of entropy in the Universe be massively decreased?*", is asked repeatedly during the course of the story. "*Insufficient data for a meaningful answer*", was always the discouraging response. But as the computer becomes big enough to permeate the entire Universe, the answer was imperative: "*Let there be light*".

If you put together a mixture of subatomic particles in a barrel, you can get rocks, or superconductors, or living organisms, or something else. But how is it possible that from the same basic matter, radically different things can emerge? This is a good problem for the Laplace devil, who, by applying scientific reductionism, assumes that complex things can be understood from the bottom level through application of the laws of physics. In other words, if you have a big computer, the prediction of the emergence of life from the prebiotic soup could be achieved. The idea that computation could be connected with physics is credited to Landauer, as mentioned in Sect. 2.4, while the calculations can be consistently undertaken with the underlying physical laws, as evidenced by Charles Bennet and Edward Fredkin. A further step has been introduced by Konrad Zuse by proposing that the Universe could be compared to a common computer made of a regular array of bits interacting with their neighbors, similarly to what occurs in a cellular automaton that consists of a regular grid of *cells*, each in one of a finite number of *states*, such as the familiar *on* and *off*.

In 2000, Seth Lloyd, professor of mechanical engineering and physics at MIT, although he refers to himself as a quantum mechanic, showed how the computational capacity of any physical system depends on the amount of available energy, as proven in the previous chapter. If we assume the Universe to be isotropic, the distance to its edge is the same in every direction, so that it can be compared to a spherical region that includes all matter that may be observed from the earth. It has the form of a sphere with volume $V = (4/3)\pi(ct)^3$ and mass $M = V \cdot \rho$, being $\rho = 10^{-26}$ kg/m^3, is the density of the present matter, t the age of the Universe and c the velocity of light. To evaluate the number of bits that can be registered by the Universe, the calculation, familiar in cosmology, of its entropy is required. This can be done by accounting for the fact that the energy, by remembering the Einstein relation, is equal to Mc^2, while the temperature can be estimated by the equation describing the behaviour of the radiation present in the cosmos. Then, by using the Eq. (2.9), it comes out that the Universe could be attributed to being about 10^{92} bits. This represents huge potential such that John Wheeler's sentence "*It from bit*" could be modified as follows: "*It from qubit*". The involvement in the calculations of the particles present in the matter implies their movement, which requires a certain amount of energy. In fact, the maximum rate at which a bit can flip, or an electron move from one state to another, depends on energy. If the inversion of a bit from up to down, or vice versa, requires the variation $\Delta\varepsilon$ of energy, Norman Margolus and Lev Levitin, mindful of the Heisenberg uncertainty principle, proposed that the transition take place in a time interval equal to $\tau = h/4\Delta\varepsilon$, h being the Planck constant. By applying this result to all of the available energy in the Universe, we obtain

$$\frac{4Mc^2}{h} = 10^{105}\,(\text{bit/s}).$$

It follows that, in its 13.8 billion years of life, the Universe may have conducted about 10^{122} operations, a huge number that defines its maximum available processing power. Such a figure establishes the limit through which the laws of physics have been operative. In conclusion, metaphorically, the Universe may be compared to a giant computer, the only one able to answer to the question: "*Can you recreate the Big Bang?*" with: "*Fiat lux!*". But what did it make with the benefit of the afore-mentioned gargantuan computing power? Let us focus our attention on the synthesis of the chemical compounds that play an essential role in the vital processes, the proteins, whose presence is essential not only for their structural properties, but also, as will see later, for their catalytic functions. Just as a reminder, they are polymers constituted by 20 types of amino acid, so that a chain with n units corresponds to 20^n isomers. A reasonable value of n for a natural protein is 200, so that the fact that there are twenty choices at each position means that the total number of possible proteins is $20^{200} = 10^{260}$ isomeric molecules. An impressive number, much higher than the bits provided by the Universe during its lifetime! This finding brings us to the term 'ergodic', which is a mixture of two Greek words: *ergon* (work) and *odos* (path), introduced by Boltzmann in statistical thermodynamics. It concerns the investigation of systems with a large number of particles in equilibrium, by assuming that, during the time taken to perform a measurement, they pass through a sequence of states that is representative of the whole set of available states. Actually, in the present version of the ergodic theory, any specific situation must be investigated, and in the case under examination, the fact that there is no way that the Universe could have created all of the possible proteins of length 200, but only a subset of them, is evidence of a strong non-ergodicity, which implies the emergence of preferential paths. Stuart Kauffman claims that it should be credited to the fact that each new combination opens the possibility of other new combinations. This is an intriguing perspective, but it tends to underestimate the rich behaviour coming from the analysis of complex networks, such as the ones that bring about the emergence of collectively autocatalytic sets. The fact that spectroscopic observations highlight the presence in the cold galactic clouds of complicated compounds from organic chemistry suggests that their synthesis occurs through the intervention of mechanisms able to generate complex molecules. Actually, in biological systems, interactions among a multitude of organic molecules are involved in reacting networks controlled by the behavior of the individual enzymes, which catalyze the reactions towards specific compounds. Typically, metabolic cycles are present in the cells by which energy is produced, as well as in those in which nucleic acids, proteins and lipids are synthetized. But what are the mechanisms through which the chemical structures are capable of incorporating the information content required to allow for the emergence of their self-reproduction?

References

Alpher R.A., H. Bethe, G. Gamow. *The origin of Chemical Elements*, Phys. Rev. Letters, 73, 7, 1948, 803-804.

Burbidge E.M., Burbidge G.R., Fowler W.A. and Hoyle Fred. *Synthesis of the Elements in Stars*, Revs. Mod. Physics 29:547–650, 1957.

Kragh Helge. *Cosmology and Controversy*, Princeton University Press, 1996. Robert Oerter. *The Theory of almost Everything*, Pi Press, New York, 2006.

Herbst Eric, Ewine F. van Dishoeck. *Complex Organic Interstellar Molecules*, Annu. Rev. Astron. Astrophys. 2009. 47:427–80.

Lloyd Seth. *Programming the Universe*, Knopf, 2006.

Lloyd Seth. *Ultimate physical limits to computation*, NATURE | VOL 406 | 31 AUGUST 2000 | www.nature.com

Miller Stanley. *Production of amino acids under possible primitive earth conditions. Science.* 117 (3046), 1953: 528–529.

Hazen Robert. *Genesis*, National Academy Press, 2005.

Chapter 4
Biomolecules, Networks and Bioenergetics: System Approach to Biology

4.1 A New Paradigm

Biology is often considered the ugly duckling of science. Compared with mathematical proofs, physical models and the molecular structures of chemistry, biology has often seemed unquantifiable and unpredictable. Despite this, from its outset and over the centuries, biology has played an irrefutable role for its implications in the safeguarding of human health. Moreover, biology contributes to the enrichment of our culture thanks to the development of the theory of evolution, and the efforts in regard to the deepest and most difficult question faced by humanity: the origin of life. With the passing of time, the different scientific disciplines have gained advantage from mutual interactions, which have, in the meantime, earned biology a hegemonic position from the cultural and applicative perspectives, by spearheading issues and technologies that are having a significant impact on the future of human society. Regrettably, this might be in a dystopic way, as was anticipated in the already-mentioned novel by Aldous Huxley, published in 1932, *Brave New World*, in which a vision of an unequal, technologically advanced future is offered, in which humans are genetically bred and pharmaceutically anesthetized to passively uphold an authoritarian ruling order.

What about Chemistry in this landscape? Some consider it the branch of the physical sciences devoted to the composition, structure, properties and change of matter. At present, it is operating within the framework of a solid reductionism anticipated by the sentence written by Paul Dirac in 1929: *"The fundamental laws necessary for the mathematical treatment of a large part of physics and the whole of chemistry are completely known and the difficulty lies only in the fact that applications of these laws leads to equations that are too complex to be solved"*. Thanks to the advancement of computer science and related technologies, it is now possible to make accurate predictions about the energy of molecules and its role in chemical reactions.

© Springer Nature Switzerland AG 2018
S. Carrà, *Stepping Stones to Synthetic Biology*, The Frontiers Collection,
https://doi.org/10.1007/978-3-319-95459-2_4

Table 4.1 Path to the prediction of physico-chemical properties

Molecular energy obtained by applying the fundamental equations of quantum mechanics, accounting for the interactions between electrons and nuclei
↓
Intermolecular forces
↓
Molecular structure and molecular dynamics
↓
Rates of chemical transformation
↓
Behaviour of chemical and biochemical systems

"To my mind, the ultimate goal of all low-energy physics is the elucidation of physico-chemical elementary processes in wave mechanical terms". This statement by the Italian-American physicist Ugo Fano sums up the attitude of those who, in the second half of the last century, were working in the emerging area between physics and chemistry, by taking advantage of the benefits brought about by quantum mechanics and the developments achieved through experimental methods.

Now, we are coming to the realization that present activities are instead mostly focused on investigations into the structures and behaviour of living matter, taking advantage of the fact that, from the mid-twentieth century, advancement has been favored by the availability of renewed experimental techniques, such as chromatography, X-ray diffraction, NMR spectroscopy, radioisotopic labelling, electron microscopy and molecular dynamic simulations (Box 4.1).

Moreover, if, a few years ago, theoretical chemists were satisfied with confirming experimental data that were already known, today, they can venture into the prediction of data useful in applicative sectors, such as biotechnologies. The results can be achieved thanks to advances in computing, which allow reasonable simulations of the structures and transformations of chemical systems to be carried out, including relatively complex ones, by following the sequence summarized in Table 4.1.

4.2   The Molecules of Life

The marriage of chemistry with biology spans approximately 400 years, although the term "biochemistry" was only introduced in 1882, in agreement with the proposal of Carl Neuberg, a German pioneer in the sector. Much of biochemistry deals with the structure and function of cellular components such as proteins, carbohydrates, lipids, nucleic acids, and other molecules. Its main focus is the understanding of how biological molecules beget the processes that occur within living cells, which, in turn, relates to the understanding of entire organisms. Among the vast number of different biomolecules, many are large polymers composed of monomeric repeating subunits; biochemistry studies their chemical properties and behaviour. The problems addressed also include protein synthesis, cell membrane transport, the behaviour of the genetic code and the way in which cells receive, process, and respond to information from the environment. All of the mentioned chemical processes are required for the survival of living organisms and are present in cells, including simple bacteria. Their occurrence implies the intervention of the molecules reported in Fig. 4.1, which make up living systems by generating a hierarchy of complex structures. Excellent descriptions of such structures can be found in different texts, but for convenience, some useful concepts are summarized in the following. To describe the situation, it may be metaphorically attractive to invoke a kind of language in which the atoms are the letters of the alphabet, the molecules are the words and the phrases correspond to the structures generated from the interactions between the different molecules. The classes of biomolecule to be found in the previous metaphor are hereafter summarized.

Proteins, as mentioned in Chap. 1, are polymers obtained from the condensation of an amino acid sequence. Each monomer has one carboxyl group (–COOH) and an amino group (–NH$_2$) in the molecule, so that the macromolecules are obtained through the elimination of a water molecule and the formation of the CONH– group, as illustrated in Fig. 1.4. The simplest arrangement that a proteinic chain can assume is helical, such as α-helix (Fig. 4.1a), as proposed by Linus Pauling and

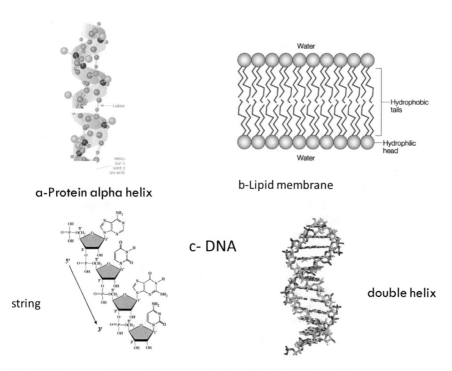

a-Protein alpha helix

b-Lipid membrane

c- DNA

string

double helix

Fig. 4.1 Structures of the most important molecules of life. For their description and analysis, see the text

Robert Corey in 1951. There are 23 naturally occurring amino acids, and it is through their combination that the variety of molecules present in living organisms is obtained. The proteins serve various functions, but above all, there are their catalytic qualities, because, as enzymes, they increase the speed of the reactions that occur in the cells.

The interpretation of the functions of the proteins proceeds from the map of the sequence of amino acids to the structure of the macromolecule, and then from the structures to their functions. The proteinic macromolecules wrap around themselves, acquiring the structure of a compact ball, but with recesses on the surface able to house the active centers needed to confer the catalytic activity.

Lipids are organic molecules characterized by a hydrocarbon tail, therefore making them insoluble in water, and a polar head, such as the carboxylic group, soluble for its acidity. For this dual characteristic, they have the tendency to self-organize by forming membranes whose structure, illustrated in Fig. 4.1b, depends only on weak intermolecular forces. A lipid membrane is a structure that forms the double-layered surface of all cells, which keeps the interior of the cell separate from the watery exterior. Their arrangements control the entry and exit of the molecules and ions involved in the cellular processes.

Carbohydrates are the most important components of the plant world, because they are present in the form of starch and cellulose, which are polymers of the general formula $(CH_2O)_n$, with n of the order of 10^4. When $n = 12$, we have the disaccharides, for instance, sucrose, whereas with $n = 6$, we have glucose, which is main product of photosynthesis.

glucose

Sugars with five carbon atoms in each molecule are called pentoses. Their relevance is due to the fact that they are the 'backbone' of nucleotides, because each side of the helix of DNA consists of a chain of pentose sugars alternating with phosphate groups. The sugar in RNA is ribose, while in DNA, it is 2-deoxyribose.

ribose (RNA)   *deoxyribose (DNA)*

DNA, or deoxyribonucleic acid, is the biopolymer constituent of genes, the repository of information that governs the cellular processes involved in the development of all living organisms. It is formed by long and thin macromolecules (Fig. 4.1c), constituted by alternating a molecule of phosphoric acid with a cyclic molecule of deoxyribose, a five-carbon sugar molecule, bound to one of the four molecules of the nitrogenous bases, abbreviated with their initial as follows:

A = adenine, T = thymine, G = guanine, C = cytosine, U = uracyl.

Adenine   Guanine

Thymine   Cytosine   Uracil

The phosphoric acid molecules follow one another along the chain through two phosphodiester bonds, each of them hooked to the deoxyribose molecule:

while the nitrogen bases occupy the position indicated by the three dots. The nitrogenous bases belonging to the two different chains are joined by means of a relatively weak hydrogen-bond formation, by coupling joints of the GC and AT types, then acting as regular supporting shelfs (Fig. 4.1c). In essence, two DNA polymer molecules interact with each other in a complementary manner by wrapping themselves according to the typical double helix structure, which has become an icon of molecular biology. Horace Freeland Judson, in his highly successful book, defined the date of the discovery of the structure of DNA by Crick and Watson as the "*The eighth day of creation*".

In essence, cellular activities are based on the co-presence of DNA that contains the recipe for their manufacture and proteins that perform the corresponding activities.

RNA, or ribonucleic acid, is instead formed by individual oligomeric chains of different length, partially wrapped on themselves. It is involved in various activities associated with the synthesis of proteins catalyzed by enzymes. While DNA contains deoxyribose without the hydroxyl group attached to the pentose ring in the second position, RNA contains ribose with a hydroxyl group, which makes RNA less stable than DNA, because it is more prone to hydrolysis. Actually, the acronym RNA indicates a ubiquitous family of large biological molecules that perform multiple vital roles in the coding, decoding and expression of genes.

Bonds   Within the afore-mentioned molecules, the atoms are held by covalent bonds that require an energy of about 80 kcal/mole to break, that is, to pull the atom apart. A particular bond, called the Hydrogen bond, is also formed when a hydrogen atom is shared by two electronegative atoms, for instance, oxygen and nitrogen in proteins.

hydrogen bonding between water and an amine

Hydrogen bonds are weaker than covalent bonds, varying in energy between 2.5 and 8 kcal/mol. The presence of networks of hydrogen bonds in proteins gives stability to the structures, while in DNA, the two strands are held together by hydrogen bonds that pair the bases in one chain with the complementary bases in the other chain. Finally, it also contributes to the energy bound to active sites of proteins playing an important role in the enzymatic catalytic mechanism.

4.3   Building the Cells

Cells are the bricks of life, since all of the chemical processes required for the survival of living organisms are present in them. Cells were first discovered and named by Robert Hooke, one of the greatest scientists of the seventeenth century and one of the key figures of the scientific revolution, in 1665. Actually, he saw under the microscope the dead walls of plant cells, specifically of cork. His description, published in a popular book entitled *Micrographia*, did not give indication of the presence of a nucleus and other organelles. The first man to witness the presence of a live cell under a microscope was Anton van Leeuwenhoek, a Dutch tradesman and scientist, best known for his work on the development and improvement of the microscope and for his subsequent contribution towards the study of microbiology. He was the first, in 1674, to see and describe a single organism, which he originally referred to as an *animalcule*.

All cells are surrounded by lipid membranes, while their interior, called cytoplasm, contains DNA and some organelles called mitochondria and ribosomes. So, what is a cell? According to Salvatore Luria it is a device for keeping the concentrations of certain given components high enough so that the chemical reactions needed for life are carried out.

Also, if all of the complex appearances of life on Earth probably share a common ancestor, the existing cells are divided into prokaryotes (size 1–10 μm) and eukaryotes (size 10–100 μm), whose characteristics have already been introduced in Chap. 1 (Fig. 1.2). The former, which lacks a membrane-bound nucleus containing DNA, involves the large domain of prokariotic microorganisms, including bacteria. Typically a few micrometers in length, bacteria have different shapes, ranging from spheres to rods and spirals, and inhabit both soil and water by living in symbiotic and parasitic relationships with plants and animals. The biochemistry of bacteria is enormously versatile, so that it has been said that what is true of bacteria is also true of elephants.

Eukaryotic cells have additional structure, because their DNA is contained in a membrane-bound nucleus, and their origin is a major evolutionary transition for which we lack information about its intermediate stages. In fact eukaryotic cells are fundamentally different from those of bacteria, starting with their physical size, so that their presence in cellular architecture forms the basis for the "prokaryotic–eukaryotic dichotomy".

Some features are due to the afore-mentioned membrane-bound organelles, which are like miniature organs, able to perform different specialized functions. Mitochondria, also if very tiny, are cellular power-plants, because, as illustrated by Nick Lane, they generate most of the adenosin tri-phosphate (ATP) that is the source of biochemical free energy through the reaction reported in Fig. 1.5. Ribosomes, contrastingly, are complex molecular machines that serve as the site of biological protein synthesis. For instance, 13 million of them are present in each cell in our livers. Living organisms include a very high number of different types of cell, because brain cells are different from blood cells, which are different from skin cells, and so on. Different cells have different features and may even have structural differences, depending on their specific purposes. In different living species, there are dozens of trillions of cells in a wide variety of forms and sizes. Almost all are invisible to the naked eye and microscopically appear as small colorless drops. The longer cells are neurons, while the biggest are female eggs. But:

- Where are the constituents of the cells built, particularly proteins and DNA?
- What are the molecular precursors that provide the pieces?
- From where does the energy needed to assemble them come?
- And finally, how are they put together?

The previous questions suggest the presence in living bodies of an ensemble of chemical reactions, called *metabolism*, which includes all chemical changes that take place within the cells, to sustain life by allowing cells to grow, develop, repair damage, and respond to environmental changes. Not by chance is the term 'metabolism' derived from the Greek word Μεταβολισμός, for "change", or "overthrow".

Metabolism represents all of the chemical changes that take place in a cell through which energy and basic components are provided for essential processes, including the breakdown of some molecules and the synthesis of new ones. All of this happens through a dense network of interactions that transform the cells into a region of overcrowded molecular traffic. Two aspects deserve a deeper look: the energy involved and the nature of the chemical transformations.

4.4 Viruses

Among the dubious entities present in the biological world, viruses occupy a prominent position. In Latin, 'virus' means 'poison', or more generally, a toxic substance. Today, more than 3700 types are known, although their origin is unknown. A virus is an infectious and parasitic agent that can only multiply in the cells of other living organisms. Viruses are much smaller than bacteria, too small to be seen with a regular microscope or to be trapped, even by ceramic filters. Each of them is incapable of independent metabolic activity and replication, so that they must invade other cells to derive the energy needed to survive. In other words, a virus is basically a gene transporter with the specific purpose of infecting another cell in order to replicate. For its replication, any given virus will only provide the genetic code, while the invaded cell provides the raw material, so that the growth of the virus disrupts the internal structure of the cell, and as a consequence, the cell usually dies. Investigations by means of electronic microscopy have demonstrated that viruses are particles that have a size from between 25 and 400 microns. They consist of genetic material, either DNA or RNA, protected by a protein shell called a capsid, created by the viral genetic material and typically self-assembled. Viruses predominantly come in two kinds of shape: rods or filaments, because of the linear array of the nucleic acid and the protein subunits, and spheres, which are actually 20-sided (icosahedral) polygons. If life were a horror movie, would viruses be vampires or zombies? In other words, are viruses alive? Of course, the answer depends on the definition of life, because if we accept that, according to the Darwinian theory, evolution must be its peculiar character, viruses are not alive. Despite this, viruses are at the changing boundary between the worlds of biology and chemistry. Therefore, in their merging and reemerging with the cellular genome, the processes that, in the course of evolution, have created the successful genetic patterns that underlie all living cells can be observed. The question is open.

4.5   Thermodynamics in Action

Free energy is the fundamental tool for the synthesis of macromolecules, because it is involved in the reshuffling of chemical bonds, or in the building of large molecules from small ones. In fact, as anticipated in Chap. 2, in living organisms, the free energy is the fuel required to supply different kinds of work, including the chemical work involved in the synthesis and biosynthesis of chemical compounds. Thermodynamics teaches us that an isolated system, unable to exchange energy and work with the surrounding environment, is subject to a spontaneous transformation towards the equilibrium state, having the maximum value of the entropy. Instead, a system in contact with an environment that keeps both temperature and pressure constant is subject to a transformation that leads to the state with the minimum value of free energy. The free energy of a mixture of components subject to a chemical reaction can be evaluated as illustrated in Boxes 4.2 and 4.3. Of course, during the transformation, the composition of the system changes, in agreement with the corresponding decrease in free energy. As to the compounds of organic chemistry, when put into contact with oxygen, they are subject to an evolution that appears trivial, because it leads to the combustion products, namely CO, CO_2, H_2O, oxides of nitrogen, phosphorus, and so on. Its accomplishment is just a matter of time. Let us focus attention on the oxidation processes that take place in living organisms by considering the transformations of ingested glucose, a carbohydrate with high contents of free energy. The process has been deeply investigated and the obtained experimental results are summarized by the cycle of reactions described in Fig. 4.2. The number of components involved in the transformation is elevated, and each of them turns into each other in a cyclic succession capable of grabbing the free energy of glucose molecules so as to carry out the work required for the survival of living organisms. The final products are carbon dioxide and water, in agreement with thermodynamics, but in the meantime, the capture of the biochemical free energy through the phosphorylation reaction is occurring through the regeneration of ATP from ADP plus phosphoric acid, the reverse of the reaction reported in Fig. 1.5. Conversely, if the glucose is put into contact with the atmospheric oxygen, at a relatively high temperature, its oxidation occurs in a short time with the dispersion of the energy into the environment in the form of heat, and then an increase in the temperature, as happens in the combustion processes. This outcome demonstrates that the cells cannot use the thermal energy to feed their metabolism, because its potential, that is, the temperature, increases the rate of the chemical reactions that leads to thermodynamic equilibrium. In fact, only if metabolic reactions are running

Fig. 4.2 Krebs, or cytric acid cycle, in which the pyruvate coming from the glycolysis is transformed through a series of chemical reactions present in all aerobic organisms, by which the energy stored in carbohydrates, fats and proteins is released as carbon dioxide and chemical energy in the form of ATP. For any one molecule of pyruvate transformed, two molecules of ATP and six of NADH are produced

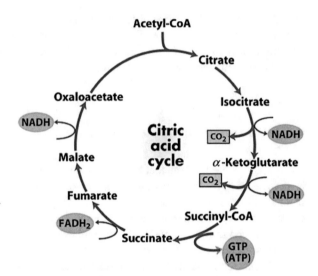

can cells stay out of equilibrium; therefore, their chemical machinery must be driven by the free energy provided by:

– **Light**, which is trapped by plants and some bacteria and then used in photosynthesis to promote well-defined metabolic transformations, which generate carbohydrates plus oxygen:

$$6CO_2 + 6H_2O + \text{solar free energy} \rightarrow C_6H_{12}O_6 + 6\,O_2.$$

The characteristics of this reaction, which produces compounds with a high content of free energy, will further detailed in Chap. 6.

– **Chemical energy** of some molecular bonds, particularly in ATP, which, if transferred to other molecules at room temperature, contributes to the occurrence of chemical reactions in metabolic pathways in which one reaction "feeds" the next.

But how does an energetically unfavorable reaction take place, as required in many transformations that occur in biology, such as the synthesis of DNA, RNA, proteins, and sugars? It must be coupled with favorable ones, so that the net free energy change is negative, as shown in Box 4.3.

At the molecular level, the exchange of free energy increases the number of available states by facilitating the synthesis of new types of molecule by contributing to the diversification of the system. In fact, while the thermal energy mainly contributes to the kinetic, vibrational and rotational motions of the molecules, the free energy instead affects the behaviour of the atomic and molecular electrons through the breaking up and formation of intramolecular bonds, as well as increasing the number of available electronic states. Therefore, it can promote the progressive development of diversified networks of chemical reactions by fostering the presence of interrelated configurations that increase the complexity of the system. In such a framework, the peculiar chemical behaviour of the cells is due to the presence on the protein surfaces of particular catalytic sites, which exhibit astonishing enzymatic properties, able to enhance the rate of some reactions. Metabolism would be impossible without enzymes, and thus life would be unsustainable. Enzymes behave like virtual ON/OFF switches, with efficient conversion to the production of specific compounds through the promotion of the activation of characteristic reaction networks. Moreover, the presence of a certain amount of specificity towards defined substrates offers some perspectives for the effective organization of the biochemical reactions present in metabolic pathways. This chance offers important perspectives for an engineering approach to enzymatic catalysis aimed at the synthesis of successfully otained compounds.

In conclusion, metabolism can be viewed as a dynamic chemical engine that converts available raw materials into energy, as well as into the molecular building blocks needed to produce biological structures and sustain cellular functions. It takes place through a network of several chemical reactions whose structure and stability is determined by the rates of the individual reactions that depend on the enzymatic system present in the cells.

4.6 The Path of Bioenergy

Life is not only about the spatial organization of particular, complex molecules; it is also a continuous chemical energy-releasing process far from-equilibrium conditions. In this framework, the utilization of free energy is achieved by means of oxidation–reduction processes, or redox reactions, able to convert it into useful work. Most transformations occur in mitochondria, previously defined as the "power-houses" of the cell, where the electrons are transferred from one molecule to another. As mentioned, mitochondria are tiny organelles with two membranes, the outer smooth and the inner convoluted into extravagant folds enclosing an area called the *matrix* (Fig. 4.3). Our main concern about their activities is focused on the transfer of electrons associated with changes in the oxidation states of two involved atoms. The electron donor is called reductant and the acceptor oxidant; the reductant and the oxidant work in pairs, and a sequence of oxidation and reduction reactions corresponds to the transfer of electrons along a chain of carriers similar to the flow of the electrical current in a wire.

In the energy trade, the most important vehicle is the familiar ATP, which, by reacting with water, is transformed into ADP plus phosphoric acid by releasing 8.01 kcal/mole of free energy. Another important system in the redox energy exchange includes the nicotin-amide dinucleotide NAD^+, whose structure contains ADP itself, as it appears below. It is reduced by accepting a couple of electrons in accordance with the following redox reaction:

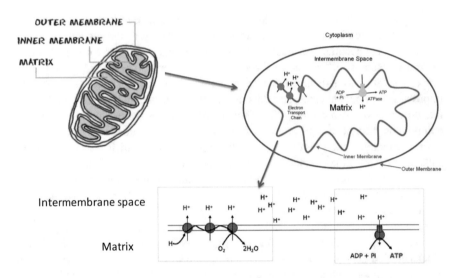

Fig. 4.3 The structure and functional behaviour of a mitochondrion, in which both its membranes, external (blue) and inner (red), are shown. The details on the path of protons in the inner membrane are also illustrated. For a detailed description of the involved processes, see the text

$$\mathrm{NAD^+ + H^+ + 2e^-} \longrightarrow \mathrm{NADH}$$

$\Delta G = 3.53$ kcal/mol > 0 for the reduction reaction, thus the reverse oxidation reaction ($\Delta G < 0$) occurs spontaneously. The reaction can then facilitate the reduction of a molecule X, by donating a proton:

$$\mathrm{X + NADH \rightarrow XH + NAD^+ + e^-}$$

The breakdown of organic matter occurring in cellular metabolism releases the free energy needed to produce other substances, such as proteins from amino acids. The processes involved are accomplished starting from a few organic compounds that come from the environment, particularly from photosynthesis, which is the basic process for feeding the free energy required for most metabolic processes. In fact, the key occurrence is the transformation of glucose, which, as previously illustrated, is the simplest carbohydrate, with the molecular formula $C_6H_{12}O_6$. It is obtained from water and carbon dioxide, using the free energy coming from sunlight, and then stored in plants. In Chap. 6, the physico-chemical mechanism of photosynthesis will be examined more deeply, including its role in the evolution of our planet and its relevance for the present and future energy economy of human society.

Now, let us again focus attention on the cell, in order to illustrate how the aforementioned sources of energy are utilized through the process called cellular respiration, which includes a set of reactions that take place to convert the biochemical free energy obtained from carbohydrates into ATP and NADH and release waste products. In the overall process, the carbon present in glucose is converted into the fully oxidized CO_2.

The first involved reactions can be summarized by the following chain:

glucose   glyeraldehyde 3-phosphate
                          G3P   pyruvate

Pyruvic acid is an important molecule involved in the synthesis of fatty acids and amino acids; it is present at the intersection of multiple biochemical pathways. As illustrated, its anion is the end product of glycolysis, which is then further transported to the mitochondria to be transformed into citric acid. The previous pathway, called glycolysis, is present in almost all organisms. In it, glucose is split into two of three carbon molecules of glyceraldehyde-3-phosphate, which are subsequently oxidized to form two molecules of pyruvic acid, the simplest organic acid of the keto functional group –CO–. Since none of the carbon atoms are oxidized to the state of carbon dioxide, little energy is released compared to complete oxidation. The further transformation of the pyruvic acid involves the intervention of the cyclic series of chemical reactions called the Krebs cycle, reported in Fig. 4.2, first written about in 1930 by Hans Krebs, a prolific biochemist who made huge contributions to the study of metabolism. A biochemical cycle is a series of reactions in which one substance joins with another, and then goes through a series of transformations by generating various products until the recovery of the original molecule. The Krebs cycle highlights the fact that in the living system, many elaborate chemical reactions are present in which one compound is changed into another and then another, through a series of connections that have inspired graphic representations that are considered icons of biochemistry. Sometimes, these are usurped for the design of decorative posters.

The Krebs cycle produces two molecules of ATP and six molecules of NADH for each molecule of pyruvate. The further final part is called the electron transport chain, and it is the only portion of the entire process of glucose metabolism in which atmospheric oxygen is involved. The process take place in a set of vast enzymatic structures, called "complexes", present in the *matrix* of mitochondria, each consisting of tens of separate proteins, each one made up of several hundreds of amino acids, weakly connected to surrounding atoms and containing multiple redox centers. In them, the electron transport occurs through a sequence of redox reactions that resemble a relay race, because they are passed rapidly from one component to the next. The entire electron transport chain, or respiratory chain, takes place in the aggregation of four of the afore-mentioned complexes, which are present in multiple copies in the inner mitochondrial membrane. The path of the electrons is illustrated in Fig. 4.3; they are taken from NAHD molecules and transferred along the chain of carriers up to the endpoint of the chain, where the molecular oxygen is involved to produce water, according to the following final reaction:

$$\text{NADH} + \text{H}^+ + (1/2)\text{O}_2 \rightarrow \text{NAD}^+ + \text{H}_2\text{O},$$

with the release of an amount of free energy equal to 3.01 kcal/mol.

Of relevance is the fact that the entire transport chain of electrons is coupled to the pumping of 10 protons, present in the matrix, across the inner membrane of the mitochondrion to its external side. Two of the membrane-bound enzymes are coupled to the process of the proton pumping through the inner mitochondrial membrane. Also, if, under normal conditions, the membrane is impermeable to protons, thanks to the presence of the afore-mentioned enzymatic complex, the pumping of protons towards the outside of the inner membrane of mitochondria takes place.

Due to its impermeability, the membrane plays the role of an electric insulator that maintains the potential difference between the inside and the outside of the inner membrane of the mitochondrion. In this situation, the machine formed by the four complexes involved in the electron chain transfer, embedded inside the membrane, employs the free energy to create the useful work required.

In the overall process, all of the carbon from the glucose is converted into fully oxidated CO_2. Mission accomplished? Not yet.

4.7 Life and Energy

In the preceding section, we understood that the electron transfer chain supplies the energy that supports the phosphorylation of ADP to ATP. But a puzzle is still present, concerning the path through which the electron transfer and the ATP synthesis are linked. In the years after the Second World War, many attempts were made to find a path that could justify its occurrence on chemical grounds, without, however, any significant results.

In 1961, Peter Mitchell, an eccentric, but creative, English scientist, introduced a radically different approach to overcoming the afore-mentioned difficulties. His unconventional hypothesis is now considered a paradigmatic transformation of the existing biochemical framework. It was baptized *chemiosmotic coupling* to suggest a process that could drive the creation of a reservoir of protons on the external side of the inner membrane of mitochondria with a resulting gradient able to do the chemical work needed for the transformation of ADP into ATP. Only in 1978 was the relevance of the previous approach officially recognised, whereupon Mitchell was awarded the Nobel Prize in Chemistry *"for his contribution to the understanding of biological energy transfer through the formulation of the chemiosmotic theory."*

As mentioned, the electron transport chain moves protons from the mitochondrial matrix, which is the space within the inner membrane, to the mitochondrial intermembrane space (Fig. 4.3). This movement decreases the concentration of the positively charged protons, resulting in a slight electrical gradient $\Delta\varphi$ across the inner membrane, called the proton-motive force (PMF); potentially, it represents a driving force for the back diffusion of the protons through the membrane.

Applying Eq. (2.2), it can be derived that the work required to transport one proton from the external side (e) to the internal side (i) of the membrane can be evaluated by means of the following expression:

$$W = \Delta G = eF\Delta\Psi + RT \ln \left(\frac{C_{H^+_{ext}}}{C_{H^+_{intt}}} \right), \qquad (4.1)$$

F being the Faraday constant and e the charge of the electron. At the equilibrium, $\Delta G = 0$, so that the value of $\Delta\Psi$ can be obtained.

Let us now assume that such a transport of protons can be associated with the phosphorylation reaction of ADP to ATP, which occurs thanks to the action of a "splendid molecular machine", as it was emphatically defined by Paul Boyer, grafted onto the membrane. In the 1960s through the 1970s, Boyer postulated that the reaction depends on a conformational change generated by the rotation of one subunit of a big asymmetric proteinic structure, indicated as a rotary-catalysis model, or ATP synthase. A research group led by the British chemist John Walker defined the catalytic characteristics of ATP synthase, confirming that Boyer's rotary-catalysis model was correct. Boyer and Walker shared the 1997 Nobel Prize in Chemistry. The emphasis on the previous definition by Boyer turned out to be altogether justified!

The subunit organization of ATP synthase and its three steps in energy transformation, illustrated in Fig. 4.4, are, respectively:

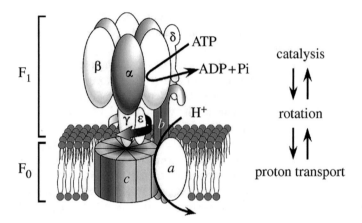

Fig. 4.4 The structure and behaviour of the ATPase, the molecular machine in which the phosphorylation reaction of ADP to ATP takes place. Its mechanochemistry combines a ratchet mechanism able to trap the Brownian fluctuations to produce a directed rotation and a power stroke due to the electrical gradient present in the membrane that, when it overcomes a given value, provides the energy that can be converted into a rotary torque in the reverse direction

(a) The chemical transformation,
(b) The mechanical rotation,
(c) The ion translocation.

Its mechanochemistry (see Yoh Wada, Sambongi and Futai), combines a ratchet mechanism, as illustrated in the first chapter, capable of trapping the Brownian fluctuations to produce a directed rotation, and a power stroke resulting from the electrical gradient present in the membrane that, when a given value is overcome, provides the energy that can be converted into a rotary torque in the reverse direction.

When the former wins, which is the more common situation, the synthesis of ATP from its constituents, ADP and phosphoric acid, takes place. When the latter prevails, ATP is hydrolyzed and the motor is reversed, turning into an ion pump that drives ions across the membrane against the electrochemical *gradient.*

Which motor "wins", by developing more torques, depends on the cellular conditions.

In conclusion, in the overall transformation of glucose into carbon dioxide, two steps in series are operative: in the former, the glycoside, ATP and NADH are produced from ADP and NAD^+, while in the latter, ATP and H_2O are produced from NADH.

One scheme of the overall process, with the energy fluxes and the balances of the free energy produced from the mentioned components, in term of ATP produced, is the following:

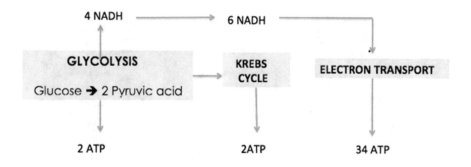

4.8 Taming the Rates

In the nineteenth century, the researchers involved in the study of the fermentation of sugar into alcohol, catalyzed by yeast, believed that the rate of the process was enhanced by a vital force present and operative only within the cells of the living organisms. Now, we are in a position to offer an interpretation of the phenomenon grounded in the interactions among the molecules involved and the dynamic of their movements, by accounting for the chance that when two molecules collide, they could react to form new ones. The bonds in the reacting molecules must break before new bonds can form and the amount of needed energy, the energy of activation, can be represented as an energy barrier to the reaction. In order to speed up the reaction, the barrier must be reduced, because if less energy is needed for the reaction, more molecules will possess enough of it to get over the barrier. Reducing the barrier is the job of *catalysts*, or of the *enzymes* in the biochemical reactions.

The term 'enzyme', from Greek ευζυμον, was coined in 1878 by the German physiologist Wilhelm Kuhnein. Early workers noted that enzymatic activity was associated with proteins, but the demonstration of their catalytic behaviour was confirmed by John Howard Northorp and Wendell Meredith Stanley, who were working in 1930 on the digestive enzymes pepsin, trypsin and chymotrypsin. A further important discovery was that the enzymes can be crystallized and that their structures can be revealed through X-ray diffractometry. This finding was made for the first time by David Chilton Phillips, in 1965, for lysozime, an enzyme that digests the coating of some bacteria. The obtained result marked the beginning of structural biology and promoted the efforts to understand how enzymes work at an atomic level. The enzymes are large proteinic biomolecules consisting of thousands of atoms. The reactions catalyzed by them include hydrolysis, isomerizations, oxidations, and electron-transfers. Their properties are extraordinary, because they are able to accelerate the catalyzed reactions by as much as-bilion-fold while operating under mild conditions. At same time, they are exquisitely selective, being capable of

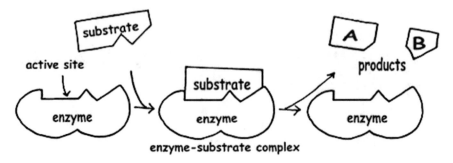

Fig. 4.5 A schematic illustration of the Lock and Key model of enzymatic catalytic reactions. The morphological correspondence between the substrate and the enzyme allows for the interpretation of the selectivity of the catalytic process

discriminating between closely related substrates, at the limit of controlling the reaction to the obtainment of a single product.

Each enzyme has a very specific job to do, as illustrated in the book of Robert Copeland, because it interacts only with the appropriate molecules, called substrates, which bond to the catalytic sites, or centers, of the enzyme (Fig. 4.5), by affecting them in such a way that makes the reaction favorable; once the reaction has occurred, the enzyme releases the products. The previous description corresponds to a standard 'lock and key' model, in which the reaction takes place through the intervention of an active site with a shape and chemical structure that is complementary to that of the substrate. Also, the X-ray diffractometry, which provides detailed information on the structure of the catalytic sites, confirms the presence of binding pockets with particular structures, able to host the substrate by affecting the value of its energy. Accordingly, two steps can be involved in the catalytic processes:

– Formation of an enzyme *(E)* plus substrate *(S)* complex *ES*:
– Transformation of the substrate *S* into the product *P*.

$$S + E \Leftrightarrow (ES) \Leftrightarrow E + P.$$

Following a classical model formulated by Leonor Michaelis and Maud Menten in 1913, the rate of the catalytic reaction, given from the number of molecules of the substrate converted per unit time by a catalytic site, can be expressed as follows:

$$rate = \frac{k_s[E_0][S]}{K_s + [S]}, \tag{4.2}$$

$[E_0]$ being the total concentration of the catalytic sites and K_s the equilibrium binding constant of the substrate with them. k_s is called the reaction rate constant, and its physical meaning will be examined more deeply later. From the previous equation, it follows that as the substrate concentration grows, the rate tends to the limiting value $\eta_{max} = k_s[E_0]$. The rate of an enzymatic reaction is also expressed as a turnover number, that is, the maximum number of chemical conversions of substrate

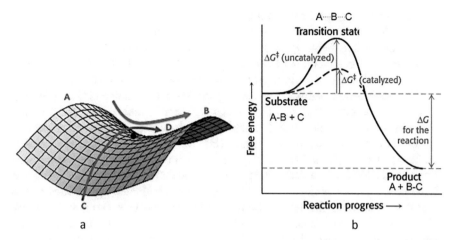

Fig. 4.6 (**a**) The black dot on the Potential Energy Surface (PES), which demonstrates a minimum along the path AB and a maximum along the path CD, corresponds to the transition state of the reaction occurring along the reaction coordinate CD. (**b**) Behaviour of the free energy change of the examined reaction along the reaction coordinate, in both the absence and the presence of a catalyst

molecules per second that a single catalytic site will execute for a given enzyme concentration.

Actually, if the spontaneous occurrence of a chemical transformation is associated with a decrease in the free energy, the rate at which the reaction approaches equilibrium also cannot be deducted from the magnitude, and the sign of its free energy changes. For instance, the oxidation of glucose is associated with a strongly negative free energy change, so that in the presence of oxygen, it is thermodynamically unstable. However, the rate of the transformation is negligible at room temperature.

To deepen this aspect, the progression of the reaction is described over a potential energy surface (PES), such as the one illustrated in Fig. 4.6 a, in which the behaviour of the energy involved in the breaking and formation of the molecular bonds that take place during the reaction is represented as a function of the geometrical configuration of the molecular reacting system, represented, by graphical necessity, only through the three components of its mass center. The path with the lowest values of the energy in going from the initial to the final state of the chemical transformation has been baptized as the reaction coordinate; as it moves along from reactants to products, an energy barrier must be overcome. The corresponding behaviour of the values of the free energy change ΔG is illustrated in Fig. 4.5b. Its maximum value $\Delta G^{\pm}$ along the path is called the "free energy of activation", and the corresponding reacting system is called the "transition state".

The presence of the energy barrier prevents the reaction occurrence, so that it can take place only if the system spontaneously increases its free energy at a level high enough to surmount the barrier. The intervention of small increments of energy at the molecular level do not violate the second law, because small fluctuations in the free energy are allowed, consistent with the average free energy decrease. Their role is similar to that of the fluctuations that give rise to the Brownian motions.

But how is it possible to evaluate the rate of an enzymatic reaction through knowledge of the structures of the reagent molecules and the active enzymatic center? This question has a long history, which started in the first half of the last century, when some scientists, Eugene Wigner, Michael Polany and Henry Eyring, were investigating the way in which a chemical transformation emerges from the dynamics of the collision between molecules, described by accounting for the existence of the potential energy surface (PES). The difficulty of the task created the feeling that the prediction of the speed of a chemical transformation was a hard nut to crack. However, a particularly effective idea emerged, based on the saddle shape of the afore-mentioned PES surface, whose maximum corresponds to the critical point at which the passage from reagents to products is taking place. A link with chemical thermodynamics emerges by assuming that the concentration of the transition state could be calculated from the activated free energy as follows:

$$\left[S^{\pm}\right] = [S]e^{-\Delta H^{\pm}/k_B T}. \tag{4.3}$$

If the states of the reacting molecular system are close to the local equilibrium along the reaction coordinate that brings it from reactants to products, the reaction rate can be expressed by the number of crossings of the energy barrier per unit time, that is, the product of $[S^{\pm}]$ times its dissociation frequency $\nu^{\pm}$. A theoretical analysis demonstrates that $\nu^{\pm} = k_B T/h$, which is a universal expression that is the same for any reaction. Then, the following expression of the reaction rate constant is derived:

$$k_{cat} = \frac{r}{[S]} = \frac{\nu^{\pm}\left[S^{\pm}\right]}{[S]} = \gamma \frac{k_B T}{h} e^{-\Delta G^{\pm}/b_B T}, \tag{4.4}$$

where γ is called the transmission coefficient, whose meaning will be further clarified later; moreover, the relevant role of the activated free energy emerges.

In 1978, R.D. Showers, following Linus Pauling's original insight, wrote, "...*the entire and sole source of catalytic power of enzymes is the stabilization of the transition state...*". This statement is as valid today as it was then. In fact, if, in the presence of an enzyme, the behaviour of the free energy along the reaction coordinate is modified by decreasing the activated free energy so that the reaction rate is increased, it follows that the understanding of enzyme catalysis implies the deepening of the effects connected with the interactions between the substrate and the enzymatic centers by accounting for the associated energetic and conformational fluctuations. In fact, the fit of the site with the substrate is poor at the beginning, but it improves at the configuration of the transition state.

The parameter γ, called the transmission coefficient, accounts for the fact that some trajectories that cross the barrier in the direction of products re-cross the dividing surface to return to the reactant region. In this case, γ is less than or equal to 1. A further effect can arise from the contribution of the quantum mechanical tunneling of protons, if they are involved in the reaction, which increases the rate by a factor higher than one. The fact, evidenced by Yuan Cha, C.J. Murray and J.P. Klinman in a paper published in 1989, is obtaining relevance in the occurence of important biochemical reactions.

Enzymes are located on large proteinic molecules consisting of thousands of atoms, and the active site may include about 100 atoms. It is stimulating to attempt interpretation of their behaviour by means of the fundamental law of the molecular physical-chemistry in accordance with the sequence introduced in Table 4.1. Unfortunately, while quantum chemical calculations are nowadays affordable for up to a few hundred atoms, the solution of the corresponding equations at the required level of accuracy is not yet feasible for an entire enzymatic system. A popular method is to describe the "region of interest" of an enzyme at the most accurate level while taking into account the surrounding proteinic environment, and possibly the solvent, by means of suitable intermolecular potentials. It is important to observe that the detailed analysis, performed up to now by means of the afore-mentioned transition state theory to enzymatic reactions (see J. Gao and D.G. Trulhar), reveals that the enhancement of the rate arises from the:

– Lowering of the free energy of activation (accelerating factor 10^{11}).
– Increasing of the transmission coefficient (accelerating factor 10^3).

As mentioned, the investigations into the metabolic system include the description of complex networks of catalytic systems, so that the ability to make predictions about the rates of the reactions involved is of great importance. In this context, the transition state theory, which was born almost a century ago, is achieving a good level of maturity, to the point of becoming an important tool in the subject, confirming the far-sighted assertion by John Hirschfelder from the 1950s, when he stated that the afore-mentioned approach provides "an epoch making concept". In fact, the results obtained by the transition state theory are so satisfactory, they give the approach the character of a heuristic tool for dealing with complex reactive systems. At present, its application is spreading in various fields concerning combustion, chemical synthesis and, of course, cell behaviour. Rather interesting is the fact that enzymes share many attributes with the molecular machines previously introduced. In fact, they bind, transform and release molecules not only for the benefit of their structure, but also in the capacity of taking advantage of the ratchet mechanism, illustrated in the first chapter, for extracting the energy required for orienting or modifying and shaping some of the molecules involved in the metabolic competition from the chaos present in the molecular environment. The case of the ATP Synthase appears to be a limit, but the capacity to steer the surrounding molecular chaos is probably present in many situations.

The behaviour of enzymes can be modified by means of particular substances, called, respectively, inhibitors and regulators. The former decreases the enzyme activity by competing with the substrate for the active site. A particular example is the so-called feedback inhibition, in which the following pathway is present:

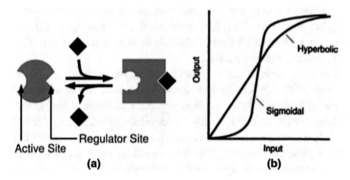

Fig. 4.7 (**a**) The non-covalent binding of a substrate with a metabolite (black square) affects its conformation towards other substrates and can positively enhance the binding with the adsorbed reactant. (**b**) Typical, either hyperbolic or sigmoidal, input-output relationships of a simple enzyme. (*Bray Dennis (1995) Protein molecules as computational elements in living cells. Nature 376:307–312*)

$$A \xrightarrow{\text{Enzyme 1}} B \xrightarrow{\text{Enzyme 2}} C \xrightarrow{\text{Enzyme 3}} D \xrightarrow{\text{Enzyme 4}} E$$

Product E inhibits Enzyme 1

where the level of E controls the activity of the enzyme A. In fact, more E is present the less active the enzyme us. The afore-mentioned is a natural example of control loops from downstream products.

Also common and of particular interest is the regulation known as allosteric, in which the regulator combines with the enzyme at a site different from the active one, as shown in Fig. 4.7. Imaginatively, some molecules are similar to the protagonists of the popular gothic novel by Robert Stevenson, Henry Jekyll and Edward Hyde, because when they bind with particular molecules, they flip from one shape to another able to catalyze a particular reaction. In some cases, the enzymes are made of several separable protein subunits, with the active center present in one of them, while the attachment site for the regulation is in a different subunit.

4.9   Towards a System Biology

The problems emerging from the previous analysis are, of course, fascinating, but at the same time, they create dismay for the underlying complexity and difficulties, both for the experimental skill required to clarify the physical and chemical nature of the involved systems and for the theoretical ability needed to provide a conceptual framework in which the mentioned aspects find an appropriate collocation.

For that which concerns the former aspect, comprehensive approaches are needed to explore the roles, relationships, and actions of the various types of molecule that make up the cells of an organism. Luckily, the presently available technologies for understanding the behavior of cells by measuring certain characteristics of their molecules, such as genes, proteins, or small metabolites, are making satisfactory progress. They have been named using the unifying suffix "omics", and include:

- **Genomics**, the study of genes and their function
- **Proteomics**, the study of proteins
- **Metabolomics**, the study of molecules involved in cellular metabolism
- **Glycomics**, the study of cellular carbohydrates
- **Lipomics**, the study of cellular lipids.

On the whole, the afore-mentioned technologies provide the tools needed to explore cellular behaviour and produce a tremendous amount of data on the functional and structural characteristics of cells. Their integration, taking advantage of the progress in computational methods, allows for the description of the behaviour of networks of interacting biological molecules in determining the properties and activities of living systems. It has been baptized "Systems Biology" for its use of mathematical descriptions of signals and circuits in the same way used for electrical engineering design and analysis. In a general sense, it is the analysis of objects in terms of interconnected functional modules with precise formal properties, thus able to contribute to the overall system performance in the frame of a mathematical description.

The approach is also making a significant contribution by accounting for the fact that the sequences of chemical reactions that control organisms could be thought of as signaling circuits similar to those used in electrical systems. Through them, cells exchange information from their surroundings via receptors present in their membranes. Inside the cell, the information is passed on to other molecules in a sequence of chemical reactions that sets up a signaling pathway able to change the cell state or the gene expression occurring within the nucleus.

In conclusion, the summary of research topics that can unify and advance the perspectives in the framework of a synthetic approach to biology are:

Measurements and extensive scouting, following the employment of the afore-mentioned technologies.

Intracellular signalling, including the mathematical modelling of the dynamical information transfer within cells.

Intercellular signalling, including the mathematical modelling of dynamical communication of information between cells within tissue and between functional biological modules.

Biological networks, focused on the analysis and modification of the complex networks that describe dynamical interactions within an organism at the biomolecular level.

The perspective is the integration of the intracellular and intercellular model components, calibrated with data from laboratory measurements into a dynamical computer-based simulation. The goal is the creation of overall computational models of the operating procedures of cellular and multicellular systems. B.O. Palsson, a pioneer in researches into synthetic biology, wrote, in 1997: *"Vast amounts of basic genetic and biochemical information are rapidly becoming available. Sequencing technology is providing us with complete information about the genetic make-up of cells".*

A few years later, he added: *"Many other fields of science and engineering have developed system science and complicated mathematical simulations. Chemicals originate from highly integrated chemical processes with structures that rival those of living cells, aircrafts are so accurately designed in computers that prototypes are no longer built. What about biology? How will the big picture emerge from a sea of biological data?"*

Some years later, in an introduction to an article in the "Economist", it was claimed that, *"A new type of biological engineering promises to speed up innovation and simplify the manufacture of drugs and other chemicals".*

In conclusion, system biology includes sophisticated experimentations and modeling for the purpose of understanding many critical processes that contribute to cellular organization and dynamics. Several recent advances in technology and in the application of models are helping in the understanding of the fundamental aspects of cell biology and, as will see, are opening interesting perspectives toward innovative applications.

Box 4.1 Experimental Techniques

Chromatography: This is employed for the separation of mixtures that are dissolved in a fluid, called the *mobile phase,* which carries the mixture through a structure holding a material, called the *stationary phase.* The separation is based on differential partitioning of the components of the mixture between the mobile and stationary phases.

X-ray diffraction: This is a non-destructive analytical technique based on the observation of the scattered intensity of an X-ray beam hitting a sample, by accounting for the incident and scattered angles, polarization, and wavelength. The obtained results reveal information about the crystal structure, the chemical composition, and the physical properties of materials and thin films.

NMR, or Nuclear magnetic resonance spectroscopy: This exploits the magnetic properties of some atomic nuclei. By relying on the phenomenon of resonance, it provides detailed information about the structure, dynamics and chemical environment of some molecules.

Isotopic labeling: This is used to track the passage of an element along a reaction path or through a metabolic pathway in a cell. The reactant is 'labeled' by replacing it with one of its isotopes, whose position is measured to follow the sequence of its displacement.

Electron microscopy: This is based on the employment of the electron microscope, which uses a beam of electrons to create an image of the specimen. It is capable of much higher magnifications and has a greater resolving power than a light microscope, allowing it to see much smaller objects in finer detail.

(continued)

Box 4.1 (continued)

Molecular dynamics (MD): This is a computer simulation method for studying the movement of atoms and molecules that are allowed to interact, giving a view of the evolution of the system. The trajectories of atoms and molecules are determined by numerically solving the Newton equations of motion for systems of interacting particles. The method, originally developed in the late 1950s, it is widely applied at present in material science and in the modelling of biomolecules.

Box 4.2 Coupling of Reactions

Let us start with the reaction that is thermodynamically unsuitable:

$$A + B \rightarrow AB, \text{with } \Delta G_{AB} > 0.$$

This can be coupled to a favorable reaction,

$$XY \rightarrow X + Y \text{ with } \Delta G_{XY} < 0,$$

through the sequence

$$A + XY \rightarrow AX + Y,$$

$$AX + B \rightarrow AB + X.$$

The result of the two reactions is

$$A + B + XY \rightarrow AB + X + Y.$$

The free energy change is the sum of the two independent reactions:

$$\Delta G_{overall} = \Delta G_{AB} + \Delta G_{XY}.$$

If $\Delta G_{overall} < 0$, the first reaction proceeds. In biochemistry, the main reaction carrier is ATP, which releases free energy through the reaction shown in Fig. 1.5.

In conclusion, the described chemical coupling is an important means by which an endergonic (or endothermic) reaction, occurring only through the absorption of heat, can take place. Additionally, its mechanical analogy is illustrated in the following figure, where, at the left, two weights attached to strings are uncoupled so that they fall in accordance with gravitational law,

(continued)

Box 4.2 (continued)

while neither of them does work on the other. When the two weights are coupled, as on the right, one weight falls and the other one rises, so that the heavy weight does the work on the light one, as it is illustrated in the book of Haynie Doad on Biological Thermodynamics.

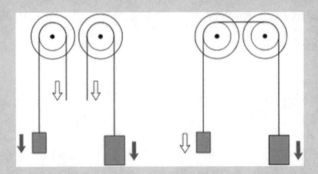

A concrete example of the above is offered by examination of the action conducted by ATP, the important source of energy in the synthesis of bio-polymers. Its involvement through coupling reactions is summarized by the following values of the numbers of the ATP molecules required per each molecular monomer added in the formation of the most important biomacromolecules.

Polysaccharides from sugar: 2

Proteins from amino acids: 4

Lipid from CH_2 unit from acetic acid: 1

DNA/RNA from nucleotide: 2

Therefore, the relevance of ATP as bio-chemical energy currency is confirmed.

Box 4.3 Free Energy Change of a Reacting Mixture

Let us refer to a mixture containing the components A, B, ... subject to the following reaction:

$$aA + bB + \ldots \Leftrightarrow mM + nM + \ldots$$

The double arrow indicates that the reaction can proceed in both directions, from left to right, or vice versa, depending on the physical conditions, which are temperature, pressure and component concentrations. In other words, the

(continued)

Box 4.3 (continued)

progress of the reaction is associated with a negative value of the free energy change ΔG associated with the reaction. It can be evaluated by means of Eq. (2.12), by utilizing (2.13) to express the chemical potential of each component. In a first approximation, let us refer to an ideal mixture by neglecting the presence of the interaction term $g_{int}(x_i)$. The following expression is derived:

$$\Delta G = \Delta G^0 + RT \ln \frac{x_M^m x_N^n \cdots}{x_A^a x_B^b \cdots} \tag{4.5}$$

being

$$\Delta G^0 = \sum_i \Delta G_i^0 (T, P = 1 atmosfere). \tag{4.6}$$

$\Delta G_i^0(T)$ is the value of the free energy of the pure component i under a standard condition, that is, at the pressure of one atmosphere and at a fixed temperature T. Their values for most of the common chemical components are known, thanks to the accurate experimental and theoretical work performed in the framework of classical thermodynamics, enriched by knowledge of the structure and dynamics of molecules. By way of example, the values of ΔG^0 for some reactions are mentioned in the sequel. In it, P_i is the phosphoric acid and the standard free energy changes are given in Kcal/mol.

$$ATP + H_2O \rightarrow ADP + P_i, \quad \Delta G^0 = -8.0,$$

$$Glucose + P_i \rightarrow glucose\text{-}6\text{-}phsphate + H_2O, \Delta G^0 = 3.3,$$

$$Ethylacetate + H_2O \rightarrow ethanol + acetate, \Delta G^0 = -4.7.$$

Photosynthesis:

$$6CO_2 + 6H_2O \rightarrow six\ carbon\ sugar + 6\ O_2, \quad \Delta G^0 = 3.3.$$

At the equilibrium, $\Delta G = 0$, so that the equilibrium condition corresponds to the following relationship among the component concentrations, expressed by their molar fraction:

$$\frac{x_M^m x_N^n \cdots}{x_A^a x_B^b \cdots} = e^{-\Delta G^0/RT} \equiv K_{eq}(T). \tag{4.7}$$

(continued)

Box 4.3 (continued)

The previous relationship, together with the stoichiometric constraints, which account for the conservation of the different atoms present in the system, allows us to evaluate its composition.

For instance, the hydrolysis reaction of ATP to ADP,

$$ATP + H_2O \Leftrightarrow ADP + P_i,$$

P_i being the phosphoric acid, is associated with a standard free energy change equal to -7.3 kcal/mol. The equilibrium condition at 25 °C can be written as follows:

$$\frac{x_{ADP}x_{P_i}}{x_{ATP}x_{H_2O}} = e^{-\Delta G^0/RT} = e^{7300/298 \times 1.98} = 2.34.10^5. \tag{4.8}$$

From the molar fractions, the values of the molar concentrations C_i (moles per volume) can be evaluated by means of the following relation:

$$C_i = \frac{x_i \rho}{\sum_i x_i M_i}, \tag{4.9}$$

ρ being the density of the mixture and M_i the molar mass of component i.

References

Luria Salvatore. *Lectures in biology*, MIT Press, Cambridge Massachusetts, 1975.

Fox Keller Evelyn. *The Century of the Gene*, Harvard University Press, 2000

Elliott, William H, Daphne C. Elliott. *Biochemistry and Molecula Biology*, Oxford University Press, 2001.

Morowitz Harold J. *Energy Flow in Biology*, Academic Press, New York, 1968.

Lane Nick. *Power, Sex, Suicide, Mithocondria and the Meaning of Life*, Oxford, 2005.

Haynie Doald T. *Biological Thermodynamics*, Cambridge, August 2006.

Yoh Wada, Yoshihiro Sambongi, Masamitsu Futai. *Biological nano motor, ATP synthase F_oF_1: from catalysis to $\gamma \varepsilon c_{10\ 12}$ subunit assembly rotation*, Biochimica et Biophysica Acta 1459 (2000) 499^505.

Meier-Schellersheim, Martin, Iain D. C. Fraser, Frederick Klauschen. *Multi-scale modeling in cell biology*, Wiley Interdiscip Rev Syst Biol Med. 2009 ; 1(1): 4-14, https://doi.org/10.1002/wsbm.33.

Yuan Cha, Christopher J.Murray, Judith P. Klinman. *Hydrogen Tunneling in Enzyme Reactions*, vol. 243, 1325-1330, 1989.

Jiali Gao, Donald G. Trulhar. *Quantum Mechanical Methods for Enzyme Kinetics*. Annu. Rev. Phys. Chem. 2002. 53:467–505.

Copeland Robert A. *Enzymes: A Practical Introduction to Structure, Mechanism, and Data Analysis*, Wiley-VCH, Inc.,2000.

Chapter 5
Making Sense of Life

5.1 An Enlightening Contribution

Erwin Schroedinger, who had the privilege of having his name attached to one of the most important equations in physics, being at the basis of quantum mechanics, was a fascinating character. Eccentric in his social behavior, he occupied a prominent position in the intellectual landscape of the first half of the previous century. The only son of a wealthy Viennese family from the last decades of the Austro-Hungarian Empire, he taught theoretical physics at Breslau and then at the ETH of Zurich, where he remained until 1927. After a brief period in Berlin, he left Germany in 1933 as a rebuke to Nazi politics, and found hospitality in Dublin at the Institute of Advanced Studies, where, in 1943, he delivered a series of public lessons entitled *"What Is Life?"*, in which some fundamental problems of biology were addressed. In those years, the subject was generating interest in the world of physics, thanks to the contributions of Max Delbruck, a German-American physicist who was launching a research program in molecular biophysics in the USA. In particular, Schroedinger was fascinated by the ongoing developments in genetics, arising through the research works performed by Gregory Mendel, an Augustinian friar whose research into inheritance was receiving widespread posthumous recognition. The enormous impact of the Schroedinger lessons led him to write a book, published in 1947 with the same title as the conferences, that aroused great interest and some perplexities. In fact, someone went so far as to say that the things already known that were mentioned in the book were trivial, whereas his original ideas were simply wrong. If so, it must be concluded that a Genius has the prerogative of making a contribution to the advancement of science even when he is mistaken. On the first page of the book, the following sentences appear: *"The large and important and very much discussed question is: How can the events in space and time that take place within the spatial boundary of a living organism be accounted for by physics and chemistry? The preliminary answer, which this little book will endeavor to expound upon and establish, can be summarized as follows: The obvious inability of present-day*

© Springer Nature Switzerland AG 2018   83
S. Carrà, *Stepping Stones to Synthetic Biology*, The Frontiers Collection,
https://doi.org/10.1007/978-3-319-95459-2_5

physics and chemistry to account for such events is no reason at all for doubting that they can be accounted for by those sciences". This preamble contributed to the increase in the number of physicists and physico-chemists who put their skills in the service of biology, with results, over time, of great relevance. While short, the book addresses different aspects of biology, especially the problems of inheritance, highlighting what may be the characteristics of a natural system capable of transmitting information. By claiming that biology had to encompass new laws of physics not previously seen in inanimate matter, he was able to anticipate the discovery of DNA. An important point in his analysis concerns the thermodynamic behavior of a living organism, such as a cell, which is taken into consideration with the intent of explaining how it can maintain its stability. In this context, it deepened the role of entropy in vital processes, as far as concerns its connection with the degree of disorder of a system. In fact, the most striking feature of life is that entropy may reduce locally while it increases globally, and therefore suggests that, because this occurs, the cell must be subjected to a negative stream of entropy, with Schroedinger literally writing: *"It is by avoiding the rapid decay into the inert state of 'equilibrium' that an organism appears so enigmatic. An organism is fed upon a negative entropy"*. In a footnote, however, it is explained that by 'negative entropy', he really meant the free energy, but unfortunately, many subsequent authors have deceptively taken "neg-entropy" as simply being entropy with a negative sign. By replacing the flow of "neg-entropy" with that of the free energy introduced by Gibbs, it turns out that the total amount of energy of a system must be separated into a part able to produce useful work and a useless part expressed by the product of the absolute temperature times the entropy. In essence, if a free energy flow is present, it can produce the work necessary to keep a living organism in a non-equilibrium state. In other words, a hypothetical Maxwell devil present in a cell could take advantage of a flow of free energy to classify the molecules, and thus prevent their degradation.

5.2 The Nature and Role of Free Energy

All energy is not equal, and a particular form, called exergy, is employed for measuring its quality, because it is able to provide the maximum work, mechanical, electrical, and chemical, that can be obtained in recovering the equilibrium conditions between a system and its environment. Exergy is thus the fuel capable of maintaining any system, such as biosphere, ecosystems and living organisms, outside of the conditions of equilibrium. If T is the temperature of the environment, while the suffix *tot* indicates the total entropy of the system plus environment, the exergy content of the system Ψ_{ex}, is given by

$$\Psi_{ex} = T\left(S_{eq}^{tot} - S^{tot}\right) = k_B T(\ln 2)\Delta I, \tag{5.1}$$

ΔI being the change of information content with respect to the equilibrium state. The passage from the second to the third member is illustrated in Box 5.1. In other words, the more a system deviates from equilibrium, the more information is needed to describe it and the greater its capacity for carrying information. It comes out that the difference between the free energy of the system and the corresponding equilibrium value is bound to the increase of its content of information. If the temperature and the pressure of the system are equal to the ones of the environment, that is, $T = T_0$ and $P = P_0$, it turns out that the exergy change identifies with the free energy change $\Delta \Psi_{Ex} \equiv \Delta G$. In this situation, of course, the only possible transformations are chemical reactions.

Any transformation has a *characteristic time*, which is the smallest interval of time in which the change of some variables can be observed. The energies contained in the system can be partitioned into *stored* energies versus *thermal* energies that reach equilibrium in a time less than the characteristic time. Stored energies are instead those that remain in a non-equilibrium distribution for a time greater than the characteristic time of the system. Focusing our attention on biological systems, according to Colin Mc Clare, the second law can be restated as follows: "*useful work is only done by a molecular system when one form of stored energy is converted into another*". In other words, thermalized energy is unavailable for making useful work, and it cannot be converted into stored energy. McClare was the first scientist to envision the potential importance of long-lived vibrational excitations in such biological molecules as proteins. He came to the idea through a process worthy of Sherlock Holmes' statement, "*When you have eliminated the impossible, whatever remains, however improbable, must be the truth.*" The major consequence of his approach arises from the introduction of time into the analysis of transformations; there is also the possibility, of course, that his restatement is restrictive, or even untrue for situations in which the thermal energy can be directed or channeled to do useful work in a cooperative system. A cell at constant temperature and pressure is subject to spontaneous evolution towards a state of equilibrium corresponding to the minimum value of its free energy. Decay can be countered by feeding it with a flow of free energy that, in this situation, identifies with the exergy, for example contained in a flux of ATP molecules that, under steady conditions, maintain the system unaltered over time, far from equilibrium. In fact, the complexity of cells depends on the transport of electric charges (electrons, ions), synthesis and biosynthesis of chemical compounds, the transport of molecules through membranes and the increase in its surface, each due to the flux of free energy. Within this framework, the randomly fluctuating energy would also be effective if a Maxwell's demon were involved in making good use of it, exactly as happens in the molecular motors with kinesin, which moves in an appointed direction without truly violating the second law, thanks to the contribution of ATP. Each single cell has its own characteristic shape, all parts of which are in constant activity. Spatially, they are partitioned into compartments by membranes, each with its own steady state, that can respond directly to external stimuli and relay signals to other compartments of the cell itself. Then, the overall steady state is a conglomeration of processes that are spatio-temporally organized, each with characteristic times that determine the way in which the system responds and develops over time.

Finally, by taking explicit account of the characteristic times, a reversible thermodynamic process merely needs to be slow enough with respect to all of the exchanges of thermal energy that involve short periods of time. It follows that high efficiencies of energy conversion can also be attained in the processes that occur rapidly, provided the equilibration is fast enough. This means that local equilibrium may be achieved for at least some biochemical reactions in the living system. For instance, the energy released in the hydrolysis of ATP is almost completely converted into mechanical energy in a molecular machine, which can cycle autonomously without equilibration with its environment. The characteristic times of some relevant biological processes are:

> *bond vibrations* 10^{-14}ps,
> *photosynthetic electron transfer* $0.01 - 1$ps,
> *ligand binding* 10^{-2} ms,
> *average enzyme catalytic turnover* 10^{-1} ms,
> *molecular motors*   1 ms,
> *proton diffusion across the cells*   10 ms,
> *generation of cells* 10^4 s,

The microsecond ms is a time equal to 10^{-3} s, while the picosecond ps is a time equal to 10^{-12} s. It can be observed that the values range from bond vibrations to cell generation by encompassing the duration of an enzymatic catalytic event, while larger times are required for protein diffusion across bacterial cells. Thus, the possibility comes to light of comparing the thermodynamic concept of "free energy" with that of "stored energy", which concerns a characteristic time interval. In conclusion, from the point of view of thermodynamics, it can be stated that a living organism is an open system able to maintain itself in an energetic state of stationary disequilibrium and able to direct a series of chemical reactions towards its synthesis. This definition reflects the energetic aspects of metabolism, but at the same time, it emphasizes the role of self-replication as a characteristic feature of living organisms, whose nature was subsequently clarified, according to the suggestions highlighted for the first time by Schroedinger himself in his famous lectures. Molecular machines are involved in different biological energy transduction processes, as in the described coupled electron transport and ATP synthesis in oxidative phosphorylation. As will be described in the next chapter, ultrafast energy transfer processes are operating within photosynthesis, in which the first step is the separation of positive and negative charges in the chlorophyll molecules of the reaction centre, which has been identified as a process that takes place in about 10^{-13} s.

5.3   The Eighth Day of Creation

"...*it therefore seems likely that the precise sequence of the basis is the code which carries the genetical information*". This sentence is taken from the article published by Francis Crick and James Watson, in a 1953 issue of Nature's magazine. In it, the structure of DNA, shown in Fig. 4.1, is officially presented for the first time, so the event must be considered as the beginning of a cultural and technological revolution. The meaning of the discovery was anticipated by Crick a few days earlier in a letter sent to his son Michael with the following words:

"*It is like a code. If you are given a set of letters you can write down the other. Now we believe that the DNA is a code. That is, the order of the bases makes one gene different from another gene, just as one page of print is different from another*". A sketch of the molecule, similar to the one in the afore-mentioned figure, was also included.

It is a common misconception that James Watson and Francis Crick discovered DNA. In reality, it was discovered decades earlier, thanks to the work of the pioneers who came before them. In the 1940s, a team of scientists led by the biologist Oswald Avery, working at Rockefeller University Hospital in New York City, showed that DNA is the molecule that carries the genetic code, the information used to build our bodies, but until the early 1950s, significant steps towards understanding the way in which it stores and uses information was not achieved.

The history of the discovery of its structure is well known, and a personal account has been given by one of the protagonists, Jim Watson, in 1968, in a book that scandalized the scientific community with its lively and easygoing tone.

Some years later, in 1979, the science writer Horace Freedland Jadson published a book entitled "*The Eighth Day of Creation*", in which the search for the structure of DNA is described as a Shakespearean tragedy in which eccentric and brilliant people, men of honor, and those of less than honor, and a heroine were involved in the hunt for a treasure whose value was indeterminable. More recently in the book "*Life's Greatest Secret*" Matthew Cobb, professor of zoology at Manchester, deepens the astonishing drama of the moment at which genetic and information technologies merged, shaping the present way of thinking.

Actually, the discovery of the structure of DNA gave extraordinary relevance to molecular biology, especially in the connection with the, apparently remote, information science that was gaining ground from the development of the communication sector. In few words, such a connection is masterfully highlighted by the following sentences by Leroy Hood and David Galas:

> The discovery of the structure of DNA transformed biology profoundly, engendering a new view of it as an information science. Two features of DNA structure account for much of its remarkable impact on science: its digital nature and its complementarity, whereby one strand of the helix binds perfectly with its partner.
>   DNA has two types of digital information:
>
> – the genes that encode proteins, which are the molecular machines of life,
> – the regulatory gene. Nature, Vol. 421, 2003.

In fact, DNA collects all of the information about the structure of biological systems, so it is a blueprint of how to build an organism that can best survive in its native environment, and how that organism might pass on that information to its progeny. With such a discovery, Schroedinger's intuition concerning the existence of a genetic code that he identified with an aperiodic crystal was confirmed.

Moreover, it must be remembered that the performance of Crick and Watson takes advantage of the work done by Erwin Chargaff, an Austrian-American biochemist from Columbia University, who experimentally determined that in DNA, the amount of one base always approximately equals the amount of a particular second base, for instance, purine versus pyrimidine. Moreover, he showed that the composition of DNA, in terms of the relative amounts of the A, C, G and T bases, varied from one species to another by adding evidence that DNA could be the genetic material.

But what about the role of the double-stranded RNA, whose structure, illustrated in Chap. 4, introduces it as a younger brother of DNA? Its discovery by Alexander Rich and David Davies by means of X-ray crystallography came only 3 years after the publication by Crick and Watson.

As a matter of fact, shortly after the shock that came about as a result of that publication, efforts to understand the way in which proteins are encoded in DNA started through the involvement of highly qualified members of the scientific community.

5.4   A Gentlemen's Club

The RNA club was a Gentlemen's partnership, founded in 1954 by James Watson, and George Gamow, the eccentric Russian-American physicist whom we already met in the third chapter in regard to the $\alpha\beta\gamma$ theory on the origin of the Universe. His entrance into the world of DNA occurred with a strange letter, handwritten, sent in July 1953 to Watson and Crick, in which he expressed interest in their discovery, triggered by some personal speculations that each organism is characterized by a long sequence of numbers that actually could be the basis for DNA.

Specifically:

> ...the animal will be a cat if Adenine is always followed by Cytosine in the DNA chain, and the characteristic of a herring is that Guanines always appear in a pair along the chain...

Gamow addressed the link between the DNA code and the proteins by indicating that the central question was how a four-digit number in the gene was translated into an amino acid alphabet in a protein. This was just the start of a discussion that involved, over time, different biologists, and some physicists and mathematicians seduced by Gamow's enthusiasm, all together in the frame of an RNA Tie club. To each of them was given a name corresponding to one of the twenty amino acids. George Gamow postulated that sets of three bases must be employed to encode the

20 standard amino acids used by living cells to build proteins, which would allow for a maximum of $4^3 = 64$ amino acids.

In 1955, Gamow published a paper in Scientific American with the significant title "*Information transfer in living cells*", in which he wrote the following sentence: "*The nucleus of a cell is a storehouse of information*"; with this sentence, he was describing the fact that the cell transfers messages that direct the construction of new identical cells.

In the meantime, Francis Crick and others insisted on the presence of a messenger to transmit genetic information from the cell nucleus to the cytoplasm, almost certainly made of ribonucleic acid (RNA). Because notable amounts of it were found in ribosomes, the sites of protein synthesis, it was assumed that RNA was the messenger. Following the concept of a genetic control on protein synthesis, in an article published in Nature in 1961, Crick and Sidney Brenner claimed that the chemical code embodied in a gene consists of groups of three bases, and that through an experiment, thanks to the use of proflavin, a bacteriostatic agent, they were able to insert or delete base pairs to create a sequence of interest. If three base pairs were added or deleted, the gene remained functional, proving that the genetic code uses sets of three nucleotide bases, each one corresponding to an amino acid. Such an experiment demonstrated the role of three DNA bases in the encoding process for the first time.

In conclusion, the code defines how sequences of triplets of nucleotide, called *codons*, specify which amino acid will be added next during protein synthesis. With some exceptions, a three-nucleotide codon in a nucleic acid sequence specifies a single amino acid. The vast majority of genes are encoded with a single scheme, which is often referred to as the standard genetic code, or simply *the* genetic code.

Many members of the RNA Tie Club achieved professional success, with several of them becoming Nobel laureates. However, the ultimate goal of understanding and deciphering the code link between nucleic acids and amino acids was achieved by Marshall Nirenberg, who was not a member of the Club. Born in New York City, Nirenberg, after earning his Ph.D. at the University of Florida, joined the National Health Institute. In 1961, together with J. H. Matthaei, he published a landmark paper on the "Proceeding of the National Academy of Science" in which they showed that a synthetic messenger RNA made of only uracils can direct protein synthesis by obtaining the first piece of the genetic code. They used a cell-free system to build a specific protein, a poly-uracyl RNA sequence (i.e., UUUUU...), by discovering that the synthesized polypeptide consisted of only the amino acid phenylalanine. They thereby deduced that the codon UUU specified the amino acid phenylalanine. In subsequent years, Nirenberg and his group deciphered the entire genetic code by matching amino acids to synthetic triplet nucleotides, discovering that there is redundancy, in that some amino acids are encoded by more than one codon and some codons are "punctuation marks" in the mRNA message. Nirenberg and his group also showed that, with few exceptions, the genetic code was universal to all life on Earth.

In 1965, the code-breaking was finished through the formulation of the Universal Genetic Code, illustrated in Fig. 5.1. It can be considered the instruction manual used

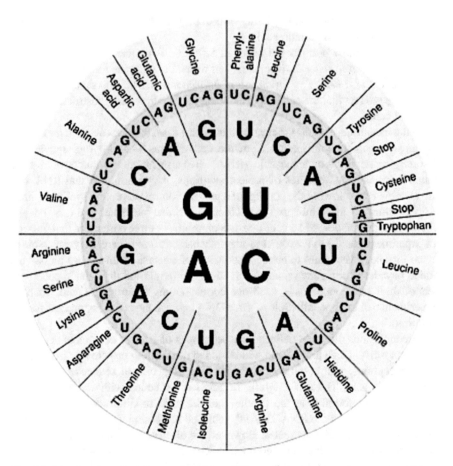

Fig. 5.1 The genetic code defines each amino acid in a protein in terms of a sequence of three nucleotides called codons. Therefore, it is the key to converting the information contained in genes into proteins. The four different nucleotides in DNA can form 64 different triple codons. Because there are only twenty amino acids in proteins, some of them are encoded by more than one codon

by cells to read the DNA sequence of a gene and build a corresponding protein, that is, as the amino acids are strung together in a chain. Each 3-letter DNA sequence, or codon, encodes a specific amino acid. The key features are:

- All protein-coding regions begin with the "start" codon, ATG.
- There are three "stop" codons that mark the end of the protein-coding region.
- Multiple codons can code for the same amino acid.

The flow of information could be visualized in simple terms as follows:

DNA ⮕ RNA ⮕ PROTEIN

Or

GENE ⮕ MESSAGGE ⮕ FUNCTION

Crick called such a flow of information "the central dogma" of biological information, valid for every existing living organism from bacteria to elephants. This is a position that has been accepted with caution, because a dogma is something that should not be disputed, and is customarily indicative of the basic belief and doctrines of a religion, not a field of science.

5.5 Central Dogma at Work

The molecular unit of heredity of a living organism, called a gene, holds the information for building and maintaining an organism's cells and passing the genetic traits on to the offspring. They are composed of DNA segments capable of managing the transmission of heredity through processes called transcription and translation, respectively, by involving the contribution of ribonucleic acid, or RNA, whose single chains are partially wrapped around themselves, as described in Sect. 4.2.

Transcription is the process by which the information contained in a DNA molecule is transferred to an adaptor molecule composed of RNA, typically 76–90 nucleotides in length, called transfer RNA (abbreviated as tRNA), locally synthesized thanks to the action of particular enzymes. It serves as the physical link between the amino acid sequence of proteins and the protein synthetic machinery of a cell.

Translation, contrastingly, refers to the process through which genetic information is transcribed onto a sequence of RNA molecules, labeled mRNA (m for messenger), complementary to that of the original DNA that acts as a mold for protein synthesis. The process takes place on a complex molecular machine, called a ribosome, upon which the involved molecular systems are placed in a row like a mounting chain, as illustrated in Fig. 5.2. The DNA-extracted message is translated into a protein chain, because each particular nitrogen-based triplet corresponds to a particular amino acid.

Ribosome is a molecular machine discovered in 1955 by George Palade, a Romanian biologist working in the USA, using the best electron microscope available at that time. It is not a simple protein, but has a more complex structure, because it is made up of about 60 proteins and three kinds of RNA, organized into two units. In the late 1960s, Harry Noler, an American biochemist, and Ada Yonath, an Israeli biochemist, through a combination of remarkable insight and a lot of perseverance, produced the first X-ray image of a ribosome, whose interpretation allowed for the identification of its structure.

The process in which a ribosome is involved, illustrated in Fig. 5.3, indicates that the involved molecular complexes interact like a couple of gears. The ribosome

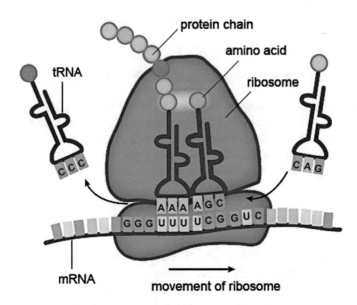

Fig. 5.2 The ribosome in action. In it, the process through which the genetic information is transcribed on a sequence of RNA molecules is complementary to that of the original DNA acting as a mold for protein synthesis. The process occurs on a complex molecular machine, in which the molecular systems involved are placed in a row, as in a mounting chain. The time course of retro-translocation has been determined from cryo-EM images of ribosomes. (*Niels Fischer NATURE| Vol 466|15 July 2010*)

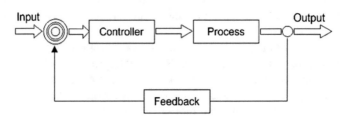

Fig. 5.3 Typical open loop feedback control system in which the output has an effect on the input quantity

grasps the tRNA strands by holding them in such a way that the codons are brought, one at a time, into a special position inside the structure of the ribosome itself. Then, it holds the strand of the messenger RNA in just the right position to bind the triplet codon of any transfer. Researches performed by means of electron cryomicroscopy, in which the sample is studied at cryogenic temperatures, provided trajectories about RNA movement through the ribosome, by demonstrating their coupling with its global and local conformational changes. The ribosome operates as a Brownian machine that couples spontaneous conformational changes driven by thermal energy to directed movement, that is, by harnessing thermal fluctuations into the directed motion of RNA. To summarize, the leading role in the overall process is exercised by ribosomes that supply the sites of protein synthesis by linking amino acids together in the order specified by the mRNA molecules. All that thanks to the contribution of its two major components, respectively, a small subunit, which reads the RNA, and a large subunit, which joins amino acids to form a polypeptide chain. Quite important is the fact that RNA, often considered a smaller brother of DNA, plays a key role in the afore-mentioned operations.

In conclusion, the genome, that is, the complete set of DNA of an organism, including all of its genes, contains the information needed to build and maintain an organism. In humans, a copy of the entire genome, including more than three billion DNA base pairs, is contained in the nuclei of all cells, so that each cell of a living organism contains the complete set of instructions for making it. The single strands of DNA are coiled up into structures called chromosomes, located in the nucleus within each cell. In this frame, life should not be thought of as a chemical event, but rather as a steady stream of information-processing. The genome is the repository of information about the world, gathered in bits over time through the process of evolution. Living organisms have been defined by Murray Gell-Mann and Jim Hartle, both theoretical physicists, as Information Gathering and Utilizing Systems (IGUS) that process information about their environment to make decisions about their actions.

5.6 Feedback Control

Without control, there could be no manufacturing, no vehicles, no industrial production activities, no computers, and so on: in short, no technology. Control systems are what make machines function as intended, and are mostly based on the principle of feedback, whereby the signal to be controlled is compared to a desired reference signal and the discrepancy is used to compute corrective control action. In other words, feedback loops enable a system to adjust its performance to meet a desired output response. Typically, an open loop feedback control system, in which the output has an effect on the input quantity, operates as illustrated in the scheme in Fig. 5.3.

The use of feedback to control a system has a fascinating history, because its first applications appeared in the development of float regulator mechanisms in Greece in the period 300 to 1 B.C. It is generally agreed that the first automatic feedback controller used in industrial processes was James Watt's flyball governor, developed in 1769, for controlling the speed of a steam engine. Its application motivated the efforts to increase the accuracy of the system through the formulation of an appropriate theory of automatic control processes, whose debut is credited to Maxwell. Actually, the exhaustive mathematical developments occurred only a couple of centuries later, during the Second World War, when it became necessary to design and construct automatic airplane pilots and other systems, all based on feedback.

Moving towards biology, cells are natural candidates to generate complex behaviour, because very large numbers of simple elements are subject to reciprocal interactions that could drive the system toward chaotic behaviour. Actually, coherent behaviour is exhibited, due to the presence of an autonomous control system constituted by large numbers of functionally diverse control elements, which interact in a selective way. Therefore, the role of feedback control in biology is widely recognized, and many scientists are working on the underlying mathematical models, which allow us to simulate and predict the performance of biological systems.

The systematic approach to the subject was begun by Norbert Wiener, the famous child prodigy and then professor at MIT, considered the originator of "cybernetics", which is a formalization of the notion of feedback with its different implications, including biology, neuroscience and the organization of society. Actually, the word "cybernetique" was first coined by the French physicist and mathematician André-Marie Ampère in his *"Essai sur la philosophie des sciences"*, published in 1834, to describe the science of civil government, including a reminiscence of the behaviour of the ancient Greek pilots. The term was borrowed by Wiener to define the study of control and communication in animals and machines. His book, *Cybernetics*, published just 4 years after the book by Schroedinger, stimulated many researchers to apply the ideas about the presence of feedback mechanisms to living organisms. At present, it is considered a seminal approach, and stands as a source of inspiration for all those who work in the field of Systems Biology. His entry into the world of biology occurred in the summer of 1961, when about two hundred biologists got together at the Cold Spring Harbor Laboratories on Long Island, New York, for a meeting devoted to Cellular Regulation Mechanisms. Its outcome is now considered an intellectual step change in the history of microbiology. The protagonists were two young French scientists, Francois Jacob and Jacques Monod, who focused attention on some peculiarities of the molecular events occurring in a cell, a breakthrough that was destined to orient most future activities in cellular biology. Thereafter, interest in microbiological issues moved from Cambridge to Paris. The important point that arose was that proteins are able to recognize more than one molecular partner by involving two binding sites. This result indicated that a proteinic enzyme could be functionally associated with a couple of processes through two sites in a way in which the activities of the two sites are completely unrelated, exactly as illustrated in Chap. 4, in which the change of shape of some enzymes was compared to the alternation within the same man of the personalities and social aptitudes respectively attributed to Doctor Jekyll and Mister Hyde.

5.7   Regulator Gene

Actually, awareness of the behaviours mentioned in the preceding section had much earlier roots, since they dated back to the 1940s, in an experiment performed on the simplest of organisms, the familiar *Escherichia coli*, which feeds on two distinct types of sugar, glucose and lactose. The former is the simplest monosaccaride, whose characteristics and structure have been illustrated in Sect. 4.2, while the latter is a disaccaride present in milk, composed of two monosaccharides, galactose and glucose. If the bacterium is fed with one of them, it quickly reproduces, growing according to an exponential curve until the nutrient is exhausted. This behavior, discovered at the beginning of the war in a sad and depressed Paris, which had been declared an "open city", attracted Monod's interest, especially because of the peculiar behavior present in the mixture of the two sugars that were consumed, or digested in succession, one after the other with only a short interruption. But how could the bacterium know that it was going from one sugar to another? And why was the glucose was digested first and the lactose second? Monod succeeded in establishing that it was due to a re-adjustment of the bacterium's metabolic system, by virtue of which, in the passage from one sugar to the other, certain specific enzymes were activated in ways suitable for the respective digestion processes. The passage required a few minutes, justifying the experimentally observed growth break. Substantially, the genes are involved in the process as molecular switches after activation, in order subsequently to determine the action of the enzymes.

In the early 1950s, Monod, together with Francois Jacob, started to analyze, in a systematic way, the activation of genes, in collaboration with Arthur Pardee, an American geneticist. Together, they discovered three principles that are at the basis of gene regulation action.

First, if a gene was turned on or off, the master copy of the DNA is kept intact in the cell. The real action manifested itself through the RNA, because after activation, a gene is induced to produce more messages, and thus more sugar digestion enzymes. Therefore, the lactose or glucose consumption does not depend on the sequence of the genes, but on the amount of RNA produced by the gene.

Second, the production of RNA messages was regulated in a coordinated manner. When the sugar source becomes lactose, the bacteria activate an entire genetic module able to digest it. Significantly, all of the genes related to a particular metabolic path are close to each other, like the books in a library, and are activated at the same time. In other words, the production of RNA messages is regulated in a coordinated manner. The sugar source becomes lactose, and the bacteria activate an entire genetic module able to digest it. In fact, the metabolic modification genetically involves the whole cell, because an entire functional gene circuit is turned on or off. All of this occurs in a manner similar to the action of a main switch in an electric circuit. For this characteristic, Monod defined such a genetic module, calling it operon.

Third, all genes have specific DNA sequences playing the role of recognition tags associated with them, the protein that can sense it, recognize it, and then activate or deactivate the genes involved. At this signal, the gene produces other RNA messages and then generates the specific enzyme to digest sugar.

In essence, the gene not only possesses the information needed to encode a protein, but it also knows when and where to fabricate it. All of that is encoded in DNA, mostly localized upstream of each gene. On the whole, the entire process can be summarized by the following circular scheme:

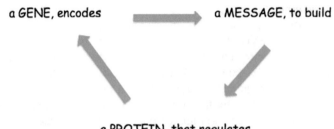

Pardee, Jacob and Monod published the results of their fundamental work in 1952, 6 years after Watson and Crick's article on DNA structure. Under the colloquial name of Paijan (from Pa-Ja-Mo), the paper quickly gained great relevance, because it highlighted that genes are not passive development projects, but rather contain a coordinated program and the means to control its execution.

Monod and Jacob were both closely associated with the bacterial geneticist André Lwoff. Even though they operated with different experimental strategies, thanks to the significant afore-mentioned contributions to the gene regulation processes, they acquired a leading position in the development of molecular biology. The Nobel Prize in Physiology or Medicine of 1965 was awarded jointly to François Jacob, André Lwoff and Jacques Monod *"for their discoveries concerning genetic control of enzyme and virus synthesis"*.

Some years later, in 1970, Monod published the book *"Le hasard e la nécessité"* (*The Chance and the Necessity*), in which he tackled themes of wide cultural relevance. This is considered the book that, after Darwin's *"Origin of Species"*, aroused the greatest scientific and philosophical debate. In it, Monod offers a general theory of living beings in which a decisive role is attributed to random initial, fortuitous mutation events that give rise to a form of life, followed by the "necessity" that translates its character according to the laws of genetic duplication.

5.8   Final Remark

Although the story of the discovery of genetic code is well-known, it is amazing to observe that the people involved in the enterprise had the most diverse educative training and were not involved in a program aimed at a well-defined purpose. Indeed:

Erwin Schroedinger, who inspired System Biology and preached the intervention of an aperiodic crystal, was a mathematical physicist,

Oswald Avery, who isolated the DNA of chromosomes, was a biologist specializing in lung diseases,

Erwin Chargaff was a biotechnologist,

Francis Crick and Jim Watson, who configured the DNA structure, were, respectively, a physicist and a zoologist,

George Gamow was a nuclear physicist.

In essence, this was one of the most heterogeneous groups of scientists of the highest quality, all driven by their curiosity. Their efforts to satisfy such curiosity changed our culture and our civilization, not just in regard to the present but also to that which has yet to come.

Box 5.1 Exergy and Information

In agreement with statistical thermodynamics, the entropy of a system is expressed by Eq. (2.1). If f_r^0 are the probabilities that maximize entropy at equilibrium, it becomes

$$S_{eq} = -k_B \sum_r f_r^0 \ln f_r^0,$$

thus, by remembering the Shannon definition of information (2.8), it follows that

$$\Psi_{ex} = T(S - S_{eq}) = k_B T \left(\sum_r f_r \ln f_r - \sum_r f_r^0 \ln f_r^0 \right) = k_B T (\ln 2) \Delta I.$$

References

Schroedinger Erwin. *What is Life?*, Cambridge University Press, 1992

Mae-Wan Ho. *The rainbow and the Worm*, World Scientific, Cambridge, 2003.

McClare, C.W.F. *Chemical machines, Maxwell's demon and living organisms*. J. Theor. Biol. 1971, 30, 1–34.

Simpson Adam, P Chris, F. Edwards. *An exergy-based framework for evaluating environmental impact*, Energy, 1442-1459, 2011.

Ridley Matt. *Francis Crick: Discover of the genetic code*, James Atlas, 2006.

Crick F. *On protein synthesis*, Symp. Soc. Exp. Biol. 12:138-163.

Gell-Mann Murray. *The Quark and the Jaguar*, Freeman, New York, 1994.

Watson James D. *DNA the secret of life*, Random House, 2003.

Monod J. Changeaux, F. Jacob, 1963. *Allosteric Proteins and cellular control systems*, J. Mol. Biol. 6.306-329.

Siddhrta Mukherjee. *The Gene*, Bobley Head, London, 2016.

Chapter 6
Complexity and Information: A Metaphor of Natural and Technological Systems

6.1 Complexity: A Philosophy or a Tool?

The challenges we face today require new ways of thinking about an interconnected and changing world, because the obtained insights will help to expand our thinking in innovative directions. The systems upon which we focus our attention are open and receive regular supplies of energy, information, and matter from the environment. Their intricacies mostly involve the presence of many elements subject to reciprocal, non-linear interactions, so that their in-depth analysis provides a framework within which the peculiar phenomena defined as emergence and self-organisation are addressed.

Emergent properties can arise from responses to environmental pressure to which the system tends to adapt. If it occurs across a population level, the system is evolving, and its behaviour concerns both science, in particular, biology, and engineering. Emergence is observed in different systems, including the interaction of molecules and the behaviour within insect societies. It reflects a universal tendency to increase their structural configuration when interactions are present among their different parts. Away from thermodynamic equilibrium, when energy flows through a collection of many interacting particles, the emergence of new patterns is facilitated so that, as will be discussed later, the energy is more easily dissipated. Unfortunately, in addition, if such phenomena surround us, a generalized mathematical treatment often remains elusive.

Also, even if it is difficult to unambiguously define what self-organisation is, there is at least agreement that it refers to a class of systems able to change their internal structure and their function in response to external occurrences. It emerges when some of their elements contribute to manipulating or connecting other elements in such a way that stabilizes either the structure or the function against external solicitations and fluctuations. Self-organization is present in galaxies and in the living world, including cells, organisms and ecosystems.

© Springer Nature Switzerland AG 2018
S. Carrà, *Stepping Stones to Synthetic Biology*, The Frontiers Collection,
https://doi.org/10.1007/978-3-319-95459-2_6

Self-organization cannot occur in isolated systems, because it needs a medium able to precondition the changes occurring within it. It is the source of instability and non-equilibrium that actuates the internal factors of the system to give rise to phase transitions. Abrupt changes in temperature, gradient of temperature, radiation, or flow of matter are specific forms of external influence.

In biological systems, the self-organising processes are often due to the search for thermodynamic stability, as happens in the formation of the lipid membranes illustrated in Fig. 4.1b. Conversely, the morphogenic model, introduced by Turing and detailed in Sect. 2.3, justifies, through the presence of retro-inhibition effects, the spontaneous formation of regular and stable structures through the creation of new forms, apparently without a detailed external blueprint. Finally, self-organization is also present in man-made systems, such as the economy and (why not?) the world of ideas.

Adaptation and evolution are processes by which the changes in the structure and behaviour of a system depend on the mutual exchange of information between it and a restless environment. It is a subject of paramount importance in biology and its performances will be more deeply examined in the next chapter.

Gottfried von Leibniz, in discussing the way in which facts that can be described by some law and those that are lawless, or irregular facts, can be distinguished, developed the simple and profound idea that a theory has to be simpler than the data it explains, otherwise it does not explain anything. This analysis is compatible with the approach to measuring complexity credited to Andrej Nikolaevič Kolmogorov, the leading Russian mathematician of the last century, and Gregory Chaitin, an Argentine-American mathematician and computer scientist. It has now been baptized "algorithmic complexity", defined as the length in bits of the shortest program that can describe an entity such as a data set, an image, a material object or a living form. Accordingly, an object without internal structure cannot be described in any meaningful way, but only by storing every feature of it. It follows that a random object has maximum complexity, since the shortest program able to reconstruct it needs to store the object itself.

In this approach, the complexity is connected to the amount of information needed to describe a system or a process. Its advantage is the computability that can be applied to describe the evolution of the systems. Moreover, it provides a framework within which the afore-mentioned phenomena of self-organisation and emergence can be addressed, so that the approach will be adopted in the following analysis. Information, according to Shannon, is expressed by Eq. (2.8). When suitable information channels are identified, what follows is mainly a matter of computation. In biological systems, the embodied channels are shaped by their interactions with the environment, so that any investigation will be an attempt to describe their time evolution.

A simple idealized version of the application of computing to complex systems is offered by cellular automata, which are discrete, spatially-extended dynamical systems, used to model many computational, physical and biological processes. Originally introduced by John von Neuman and Stan M. Ulam, they consist of regular grids called *cells*, each in one of a finite number of states, such as *on* and *off*.

An initial state is selected by assigning a state for each cell. New generations are created advancing in intervals of time, according to some fixed *rules* that determine the new state of each cell in terms of its previous state and of the states of the cells in its neighborhood. One example is offered by the "Life game", invented by J. Convay in 1970, which demonstrates the way in which, according to a set of simple rules, a complex world can be constructed in which various events can evolve. In synthesis, it develops from the following rules:

- A cell survives if two adjacent cells are alive,
- A cell dies if more than three or less than two adjacent cells are alive,
- A cell is born if three adjacent cells are alive.

Starting from an initial configuration, by means of a series of subsequent moves, the system undergoes an automatic evolution by growing and generating a sequence of configurations.

In conclusion, many invaluable applications of the knowledge focused on the afore-mentioned complex phenomena, involving science and technologies, currently exist. Our approach to their understanding will not reflect a philosophical aptitude towards observing natural systems and processes, but rather will attempt to offer some tools for carrying out experiments, making predictions, and making a computational contribution to solving certain real-world problems.

6.2   A World Patterned by Networks

Systems of interconnected or interrelated elements, called networks, are present everywhere and at every scale. As mentioned, biological cells are networks of molecules connected through biochemical reactions involved in complex metabolic schemes. At a higher scale, societies, too, are networks of people, linked by friendship, family, and professional ties, while food webs and ecosystems can be represented as networks of species. Moreover, networks pervade technology, including the Internet, power grids, and transportation systems. Network theory addresses the emergence and structural evolution of complex systems. For instance, economic complexity is organized around three major networks: the transportation network, the power network, and the communication network. These are also three components of biological complexity, involving the processing of information, energy and matter at different scales. For simple systems, we can understand the resulting behaviour intuitively, but for more complex networks, an accurate description is required, and a useful formal tool for describing and visualizing biological networks is represented by graphs, which are a set of the vertices, or nodes, with connections between unordered pairs of distinct vertices, or lines. From a mathematical point of view, the concept of the network was introduced following a logistical problem that arose in the nineteenth century in the Prussian capital, Konisberg, well-known for having given birth to Immanuel Kant. It is crossed by the river Pregel, which has a small island at its center whose access requires the crossing of some bridges. The inhabitants attempted to solve the problem of finding a route that, starting from a certain point, would went back to it only crossing each bridges once. The impossibility of finding such a path was demonstrated in 1736 by Euler, the prince of mathematicians, using a graph characterized by the presence of nodes.

In the twentieth century, two Hungarian mathematicians, Paul Erdos and Alfréd Rényi, elaborated a general model in which random networks were made up of N interconnected nodes, with k interconnections with the other nodes. If N does not change over time, and if the nodes are all equivalent, so that the tendency of each one to join another is the same, the probability $P(k)$ of a node having k interconnections follows the known Poisson distribution with a bell behaviour and a maximum. A typical example of this is the US automotive network. Contrastingly, the probability of the interconnection of airline networks does not present a maximum, but rather decreases exponentially, because there are airports called hubs, which have a very high number of connections with the others. The corresponding networks are called scale-free networks, and their probability is expressed by the following exponential law:

$$P(k) \propto k^{-\chi}, \qquad (6.1)$$

χ being an empirical parameter. The term 'scale-free network' was coined by A.L. Barabasi and his team to characterize a continuous growth and a preferential attachment, so that new nodes tend to connect to nodes that are already connected, giving rise to the evolution of the system through the addiction of nodes over time. In conclusion, a classification of the networks, as described in Fig. 6.1, can be expressed as follows:

Regular: where local groups of interconnected nodes are present.

Casual: where groups of interconnected nodes are absent and the net can be easily crossed.

Scale-free: if interconnected nodes are present, the net is also easily crossed.

Actually, hierarchical structures can also arise through originating systems that combine modularity and scale–free topology. The corresponding hierarchical models are based on the replication of small clusters linked to the central node of the original cluster. The resulting network is scale–free because it has a power–law distribution.

As mentioned, any model of cellular phenomena takes the form of interaction diagrams whose components are small molecules, macromolecules, or molecular

Fig. 6.1 Classification of the networks (see the text)

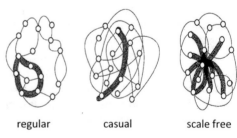

regular   casual   scale free

motors. Their interactions include a range of processes, such as chemical binding or unbinding, catalytic reactions, and regulation of activities. These processes result in the production, interconversion, transport, or consumption of the species present in the network.

Therefore, in a cell, the processes characterized by mass, energy and information transfer are connected through a complex network of cellular constituents and reactions. Any cell is built on thousands of interconnected enzymatic reactions, which control the energy fluxes and the production of cellular components involved in metabolism. The afore-mentioned exponential relation, obtained from experimental data, indicates the presence of a scale-free network. A pioneering investigation from this perspective was performed some years ago by A.L. Barabasi and coworkers, on the metabolism of *Escherichia coli*, a bacterium that, thanks to its versatility, is the most intensively studied and best understood organism on the planet. It is, in fact, a highly diverse organism with a number of complex, multi-faceted aspects. From the investigations, an elevated dishomogeneity in the distribution of the fluxes of the various components emerged, revealing a network whose composition mirrors that of a cell described just above, and therefore also indicating the presence of a scale-free network. On the whole, a "high flux backbone" of metabolism emerges, in which the various reactions are reciprocally connected to a topological structure similar to a star.

6.3   The Language of Nature

Mathematics is considered the language of the universe, so that scientists and engineers often speak of its power and elegance when describing physical reality. Starting from Galileo, the accepted language for the description of natural regularities has been mathematics, but despite its wide use, it is generally acknowledged that the high complexity of biological systems makes them incompatible with an extensive mathematical approach.

This aspect has been highlighted by John Maynard Smith and Eors Szamthmary in a book dedicated to the origin of life, through the following sentences:

> It may seem natural to think that, to understand a complex system, one must construct a model incorporating everything that one knows about the system. However sensible this procedure may seem, in biology it has repeatedly turned out to be a sterile exercise. There are two snags with it. The first is that one finishes up with a model so complicated that one cannot understand it: the point of a model is to simplify, not to confuse. The second is that if one constructs a sufficiently complex model, one can make it do anything one likes by fiddling with the parameters: a model that can predict anything predicts nothing.

All that, even if the development and extensive use of computers is enormously widening the applications of mathematics, both in the simulation of the behaviour of natural phenomena and in the investigation of particular and hidden aspects of complex systems. Nevertheless, if relatively simple calculation programs are often able to produce results rich in information, some of them also yield random results because they are extremely sensitive to initial conditions. This behaviour is common to some natural systems, such as snowflakes, that are a disconcerting mixture of regularity and casualty. All have nearly perfect hexagonal symmetry, but each one is different from another, the differences being the result of their sensitivity to external perturbations, mainly due to temperature fluctuations. Thus, it sometimes appears that the success of the physical sciences seems mainly to depend on having focused attention on the simplest things in nature, but unfortunately, such a choice seems to avoid some substantial problems whose analysis cannot be neglected. As mentioned in the previous section, an effective approach to describing the world and its evolution is to make appropriate use of the information theory, which finds its natural application in computation. Computers, the machines that manipulate information and that are currently entering into almost all aspects of our lives, ranging from budgets to atmospheric forecasts, draw their origin from the investigations into logic initiated by the afore-mentioned Gottfried von Leibnitz in the second half of the sixteenth century, who was initially known simply as Gottfried Leibniz, the *von* being added due to his great acceptance in the world of culture. Philosophers know Leibniz as a creator of an elaborate metaphysical system, but the current idea of the algorithm is present in his notebook and was rediscovered at the beginning of the last century with the shaking off of the dust of centuries.

The "algorithm", which owes its name to the Arabic mathematician Al Khwarizmi, was rediscovered by Kurt Godel, the great logician of the twentieth century, in his investigations into the computability of an analytical function through a sequence of questions and answers, articulated according to a list of instructions. This computability is said to be recursive if its value can be determined through a mechanical procedure performed by a machine by accumulating and replacing elementary components. For example, the sum of 1.234 and 5.678 is obtained by initially adding 4 and 8, which equals 12. Then, 2 is placed at the far right of the sum by remembering to add the remaining 1 to the sum of the second pair of numbers, namely 3 and 7, and so on. This recursive computability can be applied automatically by means of a machine, conceived by Alan Turing in 1936, capable of reading a finite set of instructions, called a program, each of which causes a change in the machine's status. In other words, it is an abstract machine able to manipulate symbols on a strip of tape according to a table of rules. Then, given any algorithm, a Turing machine can be constructed capable of simulating its logic content. In (3.6), it has been supposed that the evolution of the Universe is occurring because a small number of rules produces an increasingly higher complexity, starting from simple initial conditions. The Universe has been compared to a computer built by rules similar to those used by logic-based machines, including its physical systems, which are constrained by the laws of physics and chemistry. It has therefore been assumed that the Universe has evolved from simple initial conditions by attributing a role to the laws of physics parallel to the rules of an algorithm. A subtle philosophical question can arise: does the Universe, in its evolution, constantly generate new information? Or is information-processing a form of awareness exclusive to living systems? The information structures, including atoms, molecules, galaxies, stars, and planets, were major players, with an active role in everything happening in the pre-biological Universe. Natural forces controlled everything, but information was being created continuously from time zero. Its increase did not depend in any way on intelligent beings, although we are now reaping benefits from the structures inherited from pre-biological information processing.

6.4 Natural Computers

When biological systems acquire free energy from the environment, according to Eq. (5.1), in reality, they are taking in information about their environments. However, they do not simply store information, they also process it, so that the consequences of acquiring, storing, and processing information are strictly connected to their behaviour and transformations. It follows that a central goal of any approach to controlling or modifying biological systems is the design of synthetic cellular circuits able to perform complex computation and information-processing in response to specific inputs.

"*Wetware, A computer in every living cell*" is the title of a fascinating book by Dennis Bray, in which it is shown that the internal chemical behaviour of a cell is a form of computation, thanks to the presence of molecular circuits that perform logical operations. One example is offered by the behaviour of an enzyme that can adopt two different shapes, as illustrated in Fig. 4.7a. If the proteinic substrate of the enzymatic center changes shape when it binds a molecule B, it becomes able to catalyze the chemical change $A \rightarrow A'$. So, it behaves like a switch present in one of the logical operations performed in a computer program:

$$\text{If (A) and (B) then (A').}$$

This behaviour is similar to that of a transistor, which operates according to the scheme in Fig. 6.2, in which the current, passing between the connections A and A', is modulated by the voltage applied to connection B, so that its small change may give rise to a big change in A'. The previous example confirms that the ideas behind silicon computers can be transferred to biomaterials. This emerging approach has the

Fig. 6.2 Comparison of an enzyme switch associated with a protein change of shape with the typical analogy to a transistor, in which the current is modulated by the voltage applied to connection B, so that its small change may give rise to a big change in A'. (*Bray Dennis, Wetware, Yale University Press, 2009*)

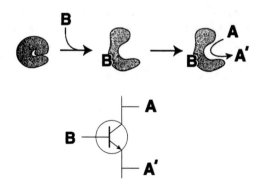

purpose of applying engineering concepts to biological systems created by a set of interacting components. Common characteristics are the components' behavior and interconnectivity, because the parts have functional, as well as structural, relationships with each other. This approach, which represents a move from molecular to modular biology, requires the modeling of the involved interactions, because the biological circuits must be translated into mathematical models that find their application in synthetic biology. Conventional computers consist of electronic circuits able to manage logical expressions built out of simple operations. Physically, the following operations are performed:

- Bits that can register 0 or 1,
- Wires that move the bits from one place to another,
- Gates that transform the bits.

The gates can have one or two inputs, but only one output, usually denoted 1 and 0. In fact, since the publication by the mathematician George Boole in 1845, it has been known that any logical expression can be built up from a set of such operations. The main operations of Boolean algebra, in fact, are the conjunction and the disjunction, and the negation, representing a formalism for describing logical relations in the same way that ordinary algebra describes numeric relations. Different logic operators can be applied to the input, and the basic types are as follows:

AND—takes two input bits and produces a single output bit equal to 1 if, and only if, both input bits are equal to 1, otherwise the output 0 is produced.

NOT—(the inverter) takes the input bits and flips it. That is, it transforms 0 into 1 and 1 into 0.

OR—takes two input bits and produces an output bit equal to 1 if one or both of the inputs are equal to 1. If both input bits are equal to 0, then it produces an output equal to 0.

NOR (OR followed by NOT)—can also be seen as an AND gate, with all of the inputs inverted.

XOR—the output is 1 if one, and only one, of the inputs is 1.

NAND—produces output 0 if one or both inputs are 1.

Their application is facilitated by the use of the tables of truth, shown in Fig. 6.3.

An example of an application is illustrated in Fig. 6.4, corresponding to an intercellular network. The system is built out of four colonies, each of which is supposed to be engineered to contain a single gate. The communications among the cell colonies correspond to the "wires" of the system, as illustrated in the figure, stimulated by the inputs mA and mB. Three cell colonies (1, 2, 3) contain NOR gates, which behave consistently with the table reported in the figure, while the entire system exhibits an XOR function.

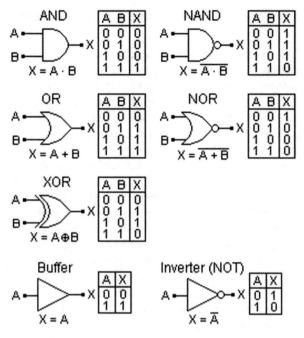

Fig. 6.3 Truth tables used to show the function of the logic gates. See the text

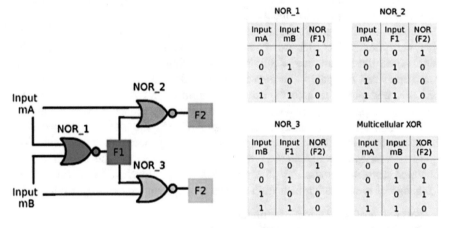

Fig. 6.4 Multicellular design of a distributed XOR circuit by conjugation of three-interacting cells. (**a**) Cell–cell communications by sender and receiver. (**b**) Truth tables for each NOR and XOR gate. (*Gerd HG Moe-Behrens, The biological microprocessor, or how to build a computer with biological parts, Volume No: 7, Issue: 8, April 2013, e201304003,* https://doi.org/10.5936/csbj.201304003)

6.5   An Abundance of Energy (and Exergy) from the Top

The fluxes of energy that come to and from the earth from the cosmos are very closed, so that our planet can be compared to a thermodynamic system under stationary conditions. Actually, the energy coming from the sun is of high quality, because it has a low entropy content, while the released energy has a high entropy content, and is therefore of low quality. It follows that local situations of thermodynamic imbalance are created, with many consequences, such as thermal energy transfer, variation of the system volume, diffusion of the chemical components and chemical reactions. Their relevance is conveniently investigated by means of the exergy function expressed by Eq. (5.1), and whose change provides the maximum work, mechanical, electrical, and chemical, that can be obtained in recovering the equilibrium conditions. Therefore, any time there is matter not in equilibrium with the environment, the potentiality of releasing free energy is present. An energy resource is made up of matter with a high exergy content, such as, for instance, an oil reservoir containing a mixture of hydrocarbons not in equilibrium with the environment in which oxygen is present. If combustion occurs, the hydrocarbons are transformed into carbon dioxide plus water, while thermal energy is released. To obtain work, the process must be coupled to suitable devices, such as an engine for performing mechanical work or a fuel cell to create electrical energy. Of course, the unavoidable presence of irreversibility decreases the amount of the obtained work with respect to its maximum value expressed from the exergy variation.

It is now worthwhile to observe that the model mentioned in 5.1 to describe the thermodynamic behavior of a living organism, if a scaling up is made, can be considered a metaphor for the behavior of our planet which, under stationary conditions, receives and radiates energy with a negative flow of entropy. Pushing this metaphor further, let us remember that Schroedinger, in his book, also highlighted the importance of solids as information carriers. Indeed, the finding that matter can compute is an amazing fact of the entire universe, because cells, bacteria, and plants technically become computers. As illustrated, our cells are continuously processing information, and the capacity of matter to compute is a precondition for life to emerge. As a matter of fact, the world is populated by structures that are always in the process of becoming more complexes. But a key question emerges: from where is the information coming? The answer can be obtained from the relationship (5.1) in which it is demonstrated that, through the exchange of exergy, the information content of a system increases by driving the:

– Formation of more complex structures;
– Promotion of autocatalytic cycling activities;
– Emergence of higher diversity.

In other words, an increasing complexity emerges from hierarchical interactions between interacting subsystems, in which the formation of structures and autocatalytic processes attains the energy necessary for maintenance, growth and development. But going back to our planet, we can agree with the following sentence credited to Shakespeare: *"It is the stars, the stars above us, that govern our conditions"*. Our planet is infused with a flux of electromagnetic energy that comes from the sun equal to 140 TW, with an amount of free energy of about 93%. Mostly, it is employed in the natural processes associated with the marine and atmospheric motions, while only 90 TW are used in photosynthesis. Ecosystems develop specialized mechanisms to harness as much solar exergy as possible, while their evolution tends to favour more complex structures, because they are more effective at utilizing free energy. In the previous framework, photosynthesis was the vehicle for transmitting the solar free energy to the biological world through its transformation into carbohydrates whose molecules, having a high free energy content, are at the basis of the food chain. In the course of roughly 4 billion years, thanks to the solar energy flow, photosynthesis has transformed the earth through the development of living organisms and the physical modification of the atmosphere through the accumulation of oxygen, which reacted with iron in the rocks, turning them a rusty red color. Therefore, it is time to look more deeply at its features.

6.6   The Emerald Planet

The vegetable world has changed the aspect and the history of our planet, not only giving it the typical emerald color of the woods, but particularly in regard to the nutrition of mankind, thanks to agriculture. The components of the plant world on which we will focus our attention are the carbohydrates, the simplest of which is glucose. According to the Italian writer and chemist Primo Levi, it is the ultimate source of our food. It follows that the solar energy harvesting reaction is one of the most important processes occurring on our planet, because it lets plants and cyanobacteria, the prokariot bacteria able to produce oxygen, absorb energy from the sun and bind it chemically through a complex metabolic process. Even the energy present in fossil fuels, for example, oil, has been stored in plants for a million years. The photosynthesis reaction can be written as follows:

$$CO_2 + 2H_2O + \text{energy (sunlight)} \rightarrow [CH_2O] + H_2O + O_2,$$

where $[CH_2O]$ is (1/6) of a sugar molecule. The previous is, of course, just the simplified synthesis of a complicated set of reactions involved in the self-conservation of living systems, occurring by virtue of the fluxes of energy and the transformations of chemical components, as summarized in the following scheme:

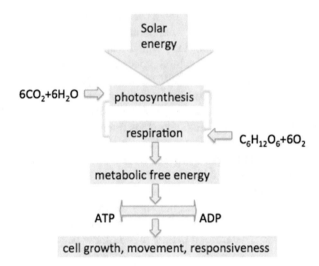

In photosynthesis, the conversion of solar radiant energy into chemical energy occurs in the presence of massive proteinic complexes. Both water splitting and carbon dioxide capture take place through the formation of carbohydrate molecules. The core of the former process is a Cubic Active Center able to reduce water according to the reaction

$$2\,H_2O + 4\,h\nu \rightarrow O_2 + 4\,H^+ + 4\,e^-.$$

The Oxygen Evolving Center (OEC) has a formula (CaO_4Mn_3), with the atoms organized in the cubic structure shown in Fig. 6.5. The researches in progress on the development of new catalytic systems for water decomposition in hydrogen and oxygen and the conversion of CO_2 into fuel require an interplay among the syntheses of new artificial systems through the adaptation of certain concepts that emerge from the understanding of photosynthesis. The overall photosynthetic process proceeds through two coupled phases, which occur in organelles called chloroplasts, present in the cells of plants.

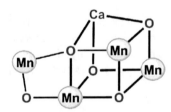

Fig. 6.5 Mn_4Ca cluster known as the oxygen-evolving complex. After four electrons have been abstracted from it, two molecules of water are oxidized to molecular oxygen, thus releasing four electrons and four protons. The photosynthetic apparatus utilizes the reducing equivalents generated to reduce CO_2 to carbohydrates

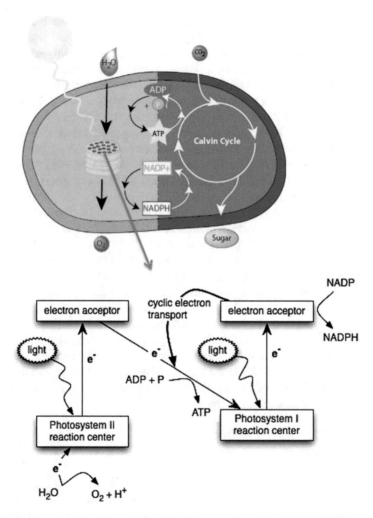

Fig. 6.6 A schematic of a chloroplast in which the connections between the different operations performed in a leaf are illustrated. The arrow indicates the reactions occurring in the Z-scheme

- **Light phase**, this takes place on a hollow membrane called the thylakoid, where the radiant energy is captured.
- **Dark phase**, this takes place in a liquid phase called stroma, present in the inner part of chloroplasts, where carbohydrate synthesis takes place.

On the whole, the entire process, as illustrated in Fig. 6.6, is split into the two afore-mentioned phases. In the former, to the left, the capture of the solar energy occurs; it is then transferred to the latter, to the right, thanks to the intervention of both of the vehicles of bioenergy transfer, which are the familiar couples ATP-ADP and $NADPH^-NADP^+$, whose redox potentials are significantly different. The excitation processes occurring in the light step are taking place because a photon flings an electron to a high energy level; its descent to a lower level transfers energy to an ATP molecule, subsequently employed to produce work in the cells. One more photon flings a second electron to a higher energy level; its descent to a lower level transfers energy to an NADPH molecule that subsequently reacts with the carbon dioxide to form carbohydrates. This description, shown in the same figure, is mainly due to the contribution of Robert Hill, a British plant biochemist, and it is known as a Z-scheme, because when the redox potentials are plotted on a graph, the electron flow looks like a zigzagging Z. On the whole, large proteins work together to excite electrons and to split water into oxygen and hydrogen ions. The illustrated light part of a photosynthetic system, which should also be considered the most important energy-transducing apparatus present in nature, looks like a machine by Rube Goldberg, the American cartoonist who became well-known for depicting complicated devices that perform simple tasks in indirect, convoluted ways, confirming that living machines are not intelligently designed and are often redundant and overly complex.

The Calvin-Benson cycle present on the right includes the series of metabolic reactions in which the carbon dioxide coming from the atmosphere and captured, thanks to the contribution of the more abundant catalyst present on the surface of the planet, called ribulose-1,5-biphosphate carbossilate or, more simply, RUBISCO, is transformed into glucose, of course, taking advantage of the energy coming from the light phase of the whole process. The net equation is

$$3CO_2 + 9ATP + 6NADPH + \text{water} \rightarrow$$
$$\text{glyceraldehyde 3-phosphate} + 8P_i + 9ADP + 6NADP.$$

The process takes place in agreement with the cycle illustrated in Fig. 6.7, whose formulation is credited to Melvin Calvin and Andrew Benson. The former was a chemist who started his scientific career at Berkeley with Gilbert Newton Lewis, the most important American chemist in the first half of the nineteenth century, by working on the employment of radio-isotopes as biological tracers. He headed a team formally named "*The Bio Organic Chemistry Group*", whose members were chemists, but during the work, biochemists were also recruited. Actually, Rubisco provides a non-optimal solution. In fact, it has a second enzymatic activity that interferes with the Calvin cycle, oxidizing one of its intermediates, the ribulose 1,5-bisphosphate. In this process, called photorespiration, oxygen is incorporated into the cycle, which undergoes additional reactions that release carbon dioxide.

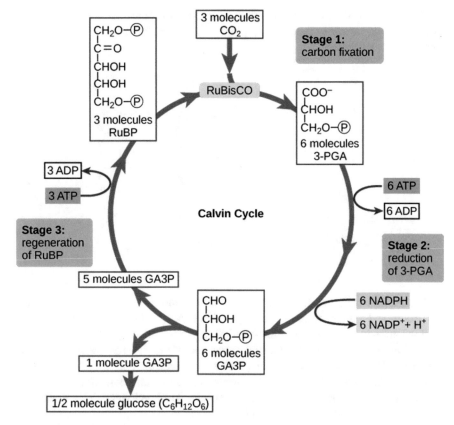

Fig. 6.7 The Calvin-Benson cycle through which glucose synthesis takes place. The cycle is the butt of chemical reactions, taking place in chloroplasts during photosynthesis. The cycle does not depend on light, because it occurs after the energy has been captured from the sun

6.7   Eating the Sun

Photosynthesis is, in appearance, a conceptually simple process: it adds a few electrons to carbon dioxide, along with a few protons to balance the charges, and then produces carbohydrates. But where are the electrons coming from? From water, by employing some of the energy coming from the sun, and thanks to the action of a catalyst that has a Cubic Active Center able to reduce water. Therefore, it is called an "Oxygen Evolving Center" (OEC, CaO_4Mn_3).

The overall process takes place through a series of reactions that can be divided into four key stages:

- Light absorption
- Charge separation
- Carbon fixation
- Oxygen evolution

The light harvesting occurs through the excitation of some natural pigments, such as chlorophyll, in whose molecules a porphyrinic group and an unsaturated hydrocarbon chain are present.

The ability to absorb light radiation is due to the unsaturation present in the molecule for the presence of alternating double bonds. Chlorophylls absorb intense light radiation with wavelengths in blue (400–500 nm) and red (650–685 nm), while absorption in the green (500–600 nm) is weaker. Therefore, the green radiation is partially reflected and gives the plants their typical emerald color.

The excitation is followed by the step of electron transfer towards a reaction center. Typically, each reaction center is connected to a light-harvesting antenna that contains 200–300 chlorophyll molecules.

Its efficiency is controlled by competition between loss through radiative and non-radiative decay and the trapping at the reaction center that takes place on a sub-picosecond timescale. At low light intensities, the efficiency of the transfer is near unity, that is, each absorbed photon almost certainly reaches the reaction center and drives a charge separation in which the energy of the impinging photon is converted into electrochemical form by increasing the redox potential difference between an oxidized donor and a reduced acceptor. At high light intensities, the efficiency decreases as a consequence of the quenching of part of the excitation energy. This makes it important to understand the design principles used by nature to improve the ability of light-harvesting systems to absorb a solar photon and transmit its energy with low dissipation. In this context, two fundamental questions must be considered:

– Is there a role for quantum mechanical effects in regard to their efficiency?
– How is the energy transport affected by the geometric characteristics of the photosynthetic protein complexes and the interactions among both electrons and electrons with nuclei?

Theoretical studies indicate that the energy coming from a photon, rather than 'hopping' from molecule to molecule, as normally assumed in energy movements, travels through multiple channels simultaneously, allowing it to pick the quickest route. In other words, the excitation is not localized on an individual molecule, but rather more than one molecule interacts with the same energy at the same time.

When the electronic interaction is relatively large, the excitation can be coherently shared among the molecules involved to form an energy–transfer funnel. This quantum wizardry enables photons to investigate multiple pathways and then choose the shortest, most efficient path, thereby leading to efficient energy transfer.

Therefore, the excitations evolve according to finely tuned interplay in the quantum superposition of the electronic states that tend to be destroyed by the coupling with the environment constituted by the protein backbone and solvent.

Quite significant is the fact that the presence of quantum coherence enables faster energy transport by demonstrating that photosynthesis relies on highly evolved molecular quantum machines. Actually, conventional wisdom suggests that quantum coherence would be quickly destroyed in large, complex, warm and wet systems, even if the afore-mentioned experiments seem to indicate that the proteinic light-harvesting complexes are tuned to capture solar energy and to transmit the excitation to the reaction centers.

Actually, due to its massive machinery, photosynthesis looks like a patchwork, typical outcome of the evolutionary bricolage, as will be examined more deeply in the next chapter. Evolution, in fact, does not necessarily result in optimum solutions, because the overall energy yield of a photosynthetic process is around a few percent. However, it does a great job, because the sun offers us a huge amount of energy, but its employment by living organisms requires adequate approaches, as will detailed in Chap. 10.

Fossil records indicate that terrestrial plants have been present on the earth only from roughly 450 million years, while the planet is at least 4.5 billion years old, and bacterial life started about 4 billion years ago. Bacteria evolved complex molecular machines able to split water by means of solar energy. The first evidence in rocks of atmospheric oxygen goes back 2.4–2.3 billion years ago; at present, there is only a procariotic group of bacteria, called cyanobacteria, that are a phylum, often called blue-green algae, able to produce oxygen.

6.8 Back to the Macroworld

As mentioned, the amount of energy that reaches the earth from the sun is largely used in natural processes involving marine and atmospheric motions, geological or chemical transformations, and, to a lesser extent, the development of biological processes. Actually, the emergence of the information potentially contained in the exergy fraction of the energy flux occurs thanks to the most amazing peculiarity of the Universe: its computing capacity. As shown, all living systems are able to process information, demonstrating the capacity to promote the emergence of new molecular structures, including the ones that are at the origin of life itself. Additionally, as matter learns to compute, it becomes selective about the information that accumulates and the structure it replicates. The networks of exergy and information exchange foster the transport movements in the cells that promote diversification, leading to self-organization at increasingly high levels of complexity, because they do not just store information, but also process it. As a consequence, the energetic capacity for biological systems to acquire, store, and process information is becoming the focus of an increasing body of interest. In human productive systems, the connections between the different activities occur through energy and information exchange networks that recall transport movements in cells, by promoting the diversification that leads to self-organization at increasingly high levels of complexity. But in order to perform a scaling up from cells to human society, it is important to realize that our world hosts people and objects whose size is orders of magnitude higher than microbiological systems. People embody the capacity to compute through the employment of their brains, while objects allow for the transmission of knowledge. In the foregoing framework, it is licit to hypothesize that socioeconomic systems have no limits for development, of course, if there is also no limit to the availability of the required resources.

Solid objects embody information and share it through the exchange of energy: mechanical, thermal, electrical and, more recently, through photon streams. The evolution of the information manifests itself in the improvement of processes through the miniaturization and better functionality of the associated technologies.

In conclusion, complexity emerges from hierarchical interactions between subsystems in which the formation of structures and autocatalytic processes attain the energy necessary for maintenance, growth and development. In other worlds, ecosystems develop specialized mechanisms to harness as much exergy as possible. Evolution tends to favour more complex structures, because they are more effective at utilizing free energy.

6.9   A Glance Towards the Future

Ray Kurzweil is a computer scientist who dedicated most of his interest to the development of technological capabilities and decreasing their cost. His attention was focused on calculus and, of course, on the development of computing equipment. Within this framework, particular relevance must be attributed to the law formulated in 1965 by Gordon Moore, cofounder of the Intel Corporation, who predicted that the number of transistors that could be placed in an integrated circuit would double every two years. The logarithm plot of the number of transistors in various commercial microprocessors against the year of their introduction confirms that the data lines up remarkably with an exponential law.

Actually, if the analysis is extended to different computing technologies, including electromechanical devices, electronic relays, vacuum tubes, transistors and integrated circuits, the plot illustrated in Fig. 6.8 is obtained. Such a finding seems

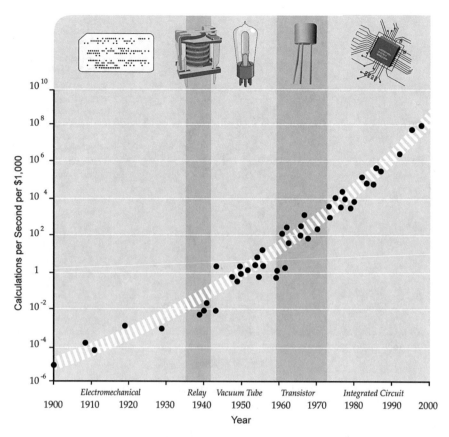

Fig. 6.8 Logarithmic plot of the Moore law concerning calculus. *From Ray Kutzweil*

to reveal a paradigmatic behaviour, reflecting an intrinsic characteristic of calculus as being a human aptitude to solve problems.

Of paramount interest, of course, is the investigation into the existence of something similar in biology, in which the techniques for DNA sequencing were subjected to rapid development, thanks to the emergence of chemical innovations and improvements in automatic technologies, so that in 1990, the governmental project of deciphering the human genome was announced. With a cost estimate of 3 billion dollars, about one dollar per base pair, the project was completed in 3 years at a cost of 2.7 billion dollars. In 2000, a parallel project headed by Craig Venter, who would go on to become one of Time's 100 most influential people, thanks to the employment of an improved approach to informational technologies, was able to sequence a second human genome in 9 months at a cost of 100 million dollars. The new technologies were comprehensive in their assorted improvements involving chemistry, automation, and, particularly, computational techniques for analysing the massive amount of data generated. Such explosive progress has enabled an increase in the availability of inexpensive DNA sequence data.

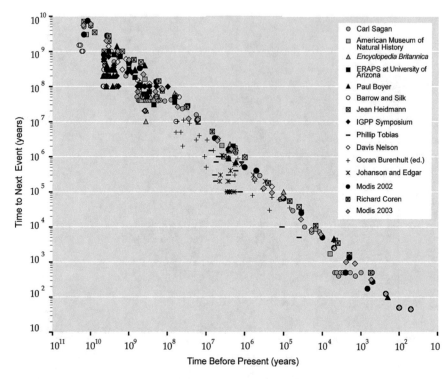

Fig. 6.9 Logarithmic plot of 15 lists of key events, starting from origin of life until the development of the personal computer. *From Ray Kutzweil*

But going back to the investigations by Kurzweil, still more impressive is the linear plot at the semilogarithmic scale in Fig. 6.9, in which the time leading up to the present is reported versus the times at which the most important cultural and technological events happened on our planet, including the origin of life. Such path-like dependence of technological development reveals that it proceeds through an evolutionary process that takes advantage of the accumulation of information, which, according to the diagram, results in being exponential. In other words, it appears that technology is, in a certain sense, a continuation of biological evolution, resulting from the imposition of the informational order on a random world. And, finally, it confirms that evolution and engineering are both iterative processes.

References

Albert-Làszlò Barabàsi. *From Network Structure to Human Dynamics*, AUGUST 2007, IEEE CONTROL SYSTEMS MAGAZINE, august, 2007, pag. 33.

Sitarba Sinha, T. Jesan, Nivedita Chatterjee. *Systems Biology: From the Cell to the Brain*, Current Trends in Science (Ed. N Mukunda), Bangalore: Indian Academy of Sciences (2009), pp 199-205.

Newman M. E. J. *The structure and function of complex networks*, http://research.compaq.com/ SRC/eachmovie/

Boccaletti, S., V. Latora, Y. Moreno, M. Chavez, D.-U. Hwang, S. Boccaletti, V. Latora, Y. Moreno, M. Chavez, D.-U. Hwang. *Complex networks: Structure and dynamics*, Physics Reports 424(2006)175–308.

Ingales, Brian P. *Mahematical Modeling in System Biology*, MIT Press, 2013. Yockey H.P. *Information Theory and Molecular Biology*, Cambridge University Press, Cambridge, 1992.

Albert-Laszo Barabasi. *Linked*, PERSEUS PUBLISHING, 2002.

Jablonsky et al. *Modeling the Calvin-Benson cycle*, BMC Systems Biology, 2011, 5:18, http:// www.biomedcentral.com/1752-0509/5/185.

Gerd HG Moe-Behrens. *The biological microprocessor, or how to build a computer with biological parts*, Volume No: 7, Issue: 8, April 2013, e201304003, https://doi.org/10.5936/csbj. 201304003.

Danchin Antoine. *Bacteria as computers making computers*, FEMS Microbiol Rev 33 (2009) 3–26.

Wolpert David H. *The Free Energy Requirements of Biological Organisms; Implications for Evolution*, Entropy 2016, 18, 138; doi:https://doi.org/10.3390/e18040138.

Morton Oliver. *Eating the Sun*, Fourt Estate, London, 2007.

Chaitin Gregory. *Thinking about Godel and Turing: Essays on Complexity*, 1970-2007, World Scientific, Singapore 2007.

Chaitin Gregory. *Randomness in Arithetic and the Decline and Fall of Reductionism in pure Mathematics*, in "Nature's Imagination", John Cornwell, Offord University Press, 1995.

Crofts, Anthony R. Life. *Information, Entropy and Time*, Complexity, pag. 14-50, © 2007 WileyPeriodicals,Inc. DOI 10.1002/cplx

Chapter 7
The Path of Evolution

7.1 The Informational Aspect of Evolution

In 1859, the English naturalist Charles Darwin explained in his book, entitled "*On the Origin of Species by Means of Natural Selection*", how biological development transformed the earliest lifeforms on Earth into the rich panoply of life seen today. He summarized his findings in the following sentence:

> Thus, from the war of nature, from famine and death, the most exalted object which we are capable of conceiving, namely, the production of the higher animals, directly follows. There is grandeur in this view of life, with its several powers, having been originally breathed into a few forms or into one; and that, whilst this planet has gone cycling on according to the fixed law of gravity, from so simple a beginning endless forms most beautiful and most wonderful have been, and are being, evolved.

Subsequently, it was estimated that such an evolutionary process has been, up to now, at work for at least 3.5 billion years.

A few years after the publication of the afore-mentioned book, in a lecture on the second law of thermodynamics, Ludwig Boltzmann made a remarkable statement about evolution and biology, by stressing the role of Darwin's work:

> If you ask me about my innermost conviction whether our century will be called the century of iron or the century of steam or electricity, I answer without hesitation: It will be called the century of the mechanical view of Nature, the century of Darwin.

The story of the discovery of evolution by means of natural selection has been told many times. Despite the fact that scientists continue to debate the relative importance of the different elements of Darwinian theory, there is nonetheless widespread agreement on its basic factors. The most intriguing aspect is the presence in the living world of a tremendous diversity within which every species must be located on a scale of increasing complexity, thus confirming that evolution is the most creative enterprise that has ever existed.

But how to approach the problems involved? "*What is true in physics is equally true in biology*", noted Ernst Mayr, a leading evolutionary biologist of the last

© Springer Nature Switzerland AG 2018
S. Carrà, *Stepping Stones to Synthetic Biology*, The Frontiers Collection,
https://doi.org/10.1007/978-3-319-95459-2_7

century, encouraging attempts to explain the diversity of living things on mathematical grounds. Actually, the employment of mathematics, as far as it concerns statistical methods, was already in progress from the first half of the past century through the application of equations able to formulate solutions to the problems concerning the statistical analysis of inheritance, species variation and their selection. Shortly after the formulation of the information theory by Shannon in 1948, it was thought that it could explain the evolution of living organisms within the frame of a new approach that could include metabolism, growth, and differentiation. Unfortunately, the first attempts, summarized in a book by H. Quaster, "*Information in Biology*", published in 1953, were not successful, because they did not offer a framework that would allow for the understanding of the broad macro-evolutionary process as we can now observe everywhere in the biosphere and in the fossil record. Moreover, the connections between the simpler and more complex forms of life were missing. This is odd, because, even though the afore-mentioned book was published in the same year as the discovery of the structure of DNA, it confirmed that the biological community was not yet ready for a quantitative description of evolutionary processes. It took almost half a century for them to realize that information is a key concept in evolutionary biology, because it unifies the digital with the biochemical approaches. In fact, when a genome is stored in a biological organism, it is involved in generating, maintaining and controlling it. But, overall, information is subject to evolution. When a population adapts to a local environment, the corresponding information is fixed in a representative genome, but as the environment changes, the information is processed and then evolves, so that it plays a relevant role in the investigations into evolutionary processes.

7.2   Evolution and Complexity

Is complexity increasing in evolution? As evidenced by Timoty Shanahan, this is a controversial question of biology, because some suggest that complexity has increased, some claim the absence of any evidence supporting it, and some deny the presence of any orienting force in evolution at all. Actually, living things increase their differentiation over time, while the emergence of new properties and the tendency towards self-organization are present, revealing the existence of a progress characterized by an increase in capacities and potentialities. This finding suggests the presence of a direction in evolution, consistent with some sort of progress, as confirmed by the history of life on the planet. Also, if many reversals have occurred along the way, on average, the history of life has moved from the simple to the more complex, in accordance with the common intuitive standard. Thus, such progress seems to be a property of evolution, including the acquisition of goals and intentions in the behavior of living organisms. Therefore, it appears to be nonsense to judge it as irrelevant. Ernst Mayr agreed, as can be ascertained through the sentence: "*By almost any measure one can think of, a squid, a social bee, or a primate is more progressive than a prokaryote*". In conclusion, the progress in evolution seems to be undeniable, like that associated with the development of a technology, due to the presence of a trend parallel to the increase in complexity. What about the opinion of the founder? Darwin's view of progress in evolution has presented something of a puzzle. Two statements, which he made at different times in his life, deserve attention. The first was written in his Notebook in 1838: "*It is absurd to talk of one animal being higher than another*." The second statement is from the "*Origin of Species*", published 20 years later: "*The inhabitants of each successive period in the world's history have beaten their predecessors in the race for life, and are, in so far, higher in the scale of nature*". Thus, if Darwin's view was occasionally enigmatic, by the end, he did not show any specific hesitation in describing the evolutionary process as progressive, as it appears in other writings before the publication of his fundamental book. The basic idea introduced by Darwin is that the evolutionary process takes place via an algorithmic procedure occurring through mutations, followed by the selection of a new generation, and then repeating the approach using the same procedure. In the present applications of the theory of evolution, called neo-Darwinism, it is assumed that species evolve by natural selection acting on genetic variations, after which the same process is repeated. But how does the complexity of biological systems emerge in a world dominated by the laws of nature, which are simple and universal, without developing a focus on well-defined outcomes? Cells transform chemical energy into the mechanical energy required for their motions and transformations along their vital cycle, but the evolution of a cell starting from defined initial conditions cannot be explained by means of specific teleological rules. Nevertheless, the information characterizing the cell's behaviour cannot be due to a random coming together of bits, because far from equilibrium conditions, there is a multiplicity of dynamic possibilities that do not leave room for purposiveness. In other words, a sort of computational strategy that underlines biological evolution must be operative. In fact, in the framework of neo-Darwinism, while mutations are random, selection occurs through the identification of particular combinations of information, which survive by taking advantage of an instruction-based behaviour, shaped through repeated selections. In other words, the existing biological systems are involved in an evolutionary process in which their molecular structures are built from the previous ones, since the instructions driving the evolutionary processes operate according an iterative computational process, such as the following one:

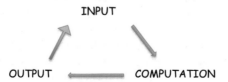

From this, a general strategy can be identified that occurs through a cyclic course that includes data and their transformations triggered by random changes. A mechanism is present that is able to copy the DNA molecules by allowing for a small number of changes, occurring, for instance, through errors, which allow for the presence of a set of outputs. The associated iterative computation takes place thanks to the intervention of a discerning process by which the information extracted from whatever random changes are made is accumulated, retaining that which better satisfies an overall selection criterion. The corresponding accumulation of information can be called "purposeful". The previous description, which shows the way in which evolution should be considered an information-processing strategy, was introduced in 1995 by Daniel Dennett, who defined it as an evolutionary algorithm, consistent with the powerful (or "dangerous") idea of Darwin that put our vision of life on a new foundation. Richard Dawkins, in turn, emphasized the role played by replicators who make copies of themselves.

Specifically, the evolution of living organisms occurs according to an operational procedure in which the DNA molecules encode the information, through the afore-mentioned iterative computational cycle by which the heritage of mutations produced in the offspring is accumulated. Nevertheless, the presence of a mechanism capable of copying the DNA allows, mainly through errors, for the occurrence of a small number of changes. In this framework, the evolution by natural selection, while not constituting a designed device, acts as if it were, because it includes a series of processes that identify the reasons for which things must be organized in a particular way.

As suggested by John Mayfield, in a meaningful title book "The Engine of Complexity, Evolution as Computation", the previous scheme can be considered a particular case of a more general theory of evolution, because it can operate in any system capable of carrying out computation.

7.3   Simulating Evolution

Let us then face the problem of simulating the evolutionary process by means of computers, by taking advantage of the afore-mentioned connections with information theory. Accordingly, it is convenient to start by storing some of the relevant information in a given sequence. Of course, attention will be focused on the genome, which is the depository of the hereditarian process, by including its interaction with the environment, which is the niche in which it survives and replicates. In a genome, a sequence of indicated with i, is present, and each of them can take on one of the four bases of DNA, with different probabilities: $\{p_C(i), p_G(i), p_A(i), p_T(i)\}$, where C stands for cytosine, G for guanine, A for adenine, and T for thymine. As illustrated in Chap. 2, the information on a system can be evaluated by means of Eq. (2.8). If it is applied to a sequence representing a genome, it quantifies its genetic identity when it is randomly selected from a pool. The information about the sequence of its elements distributed in the N sites, each with the previous probabilities, is calculated by means of the afore-mentioned equation, in which the sum goes over the different kind of bases present in the ensemble. Of course, the influence of the selection is critical to the outcome of the evolutionary algorithm, because if it is not present, all of the sequences remain equally probable, given that in each step, no one acquired an advantage over another. In the presence of selection, the population is non-uniform, because most sequences never occur, and thus do not appear, or because their fitness in the particular environment vanishes, while a few sequences are otherwise over-represented. In fact, sites are not independent, and the probability of finding a certain base at one position depends on the probability of finding another base at another position. Such correlations between sites, called epistatic, pertain to the interaction of the genes with two or more locations. Then, epistasis can significantly modify the distribution of the molecules from the one corresponding to the absence of interactions, so that the correlations are crucial in the occurrence and description of an evolutionary process, because it provides a conceptual and valuable framework for describing the time and mode of its evolution, including if a hurdle is lurking due to the presence in the genome of so-called junk DNA. Typically, the human genetic blueprint consists of 3.42 billion nucleotides, but about 98.8% of them do not code for proteins and consist of repeated transportable segments scattered randomly throughout the genome. In other words, a human being is just a very large piece of software with 6×10^9 bits $\approx$ one gigabyte, patched together and modified over the course of more than a billion years: thus, a tremendous mess. Of the many kinds of DNA sequence, some have biological functions, which are transcription and translation, and some don't do anything. But how does the described situation affect the ability of the organism to evolve? "Evolution is messy, incomplete and inefficient" is the sustaining assertion of some of its detractors. More properly, Francois Jacob asserted: "Nature is a cobbler, a tinkerer". So, evolution is a bricolage, because it does not work toward a best possible outcome, it can only do the best with what it already has. In other words, if we could design a human being from scratch, we could (probably) do a better job. Evolution only makes small changes, incremental patches, to adapt the existing situation to new environments. Metaphorically speaking, by putting together the devil of Maxwell, the bricolage of Jacob and the message of Monod, it can be concluded that evolution, through the interaction of chance due to molecular storms and the necessity imposed by physical laws, is operating as a ratchet able to rectify the random input of mutation to work towards different and renewed structures.

7.4   Evolution in a Computer

But how can a digital evolutionary program be built that would be able to perform virtual experiments with self-replicating and evolving elements? Its availability would allow us to conduct investigations on the evolution of bio-complexity, while a fallout in regard to engineering could be its application towards forecasting the evolution of some technological targets. Significant work in this area has been done by John Holland, a pioneer in the exploration of the emergent phenomena, with the intent to mimic living things evolving adaptively, by starting from the following belief: "*Living organisms are consummate problem solvers. They exhibit a versatility that puts the best computer programs to shame*". His computation metaphor relies on the model of a walk across a rugged landscape, where, in the case of biological evolution, the hills and the valleys not only depend on the properties of one species, but also on the behaviour of rival organisms.

Further progress has been made when the problem has been approached within the frame of research into artificial life, by focusing attention on communities of digital codes, each of them representing an organism forced to try out new strategies for copying and learning how to find resources on the network quickly.

An interesting example is offered by the platform called Avida, designed and built by Christian Adami, Charles Ofra and Claus Wilke, that hosts a population of self-replicating computer programs, known as avidians, whose evolution in a noisy environment is defined by the sequence of instructions that characterize their genome.

According to neo-Darwinism, the mutations resulting from random copying errors in DNA cause variations within a population of individual organisms, while natural selection acts upon these variations. Then, three basic ingredients are present in the evolutionary algorithm, respectively:

Replication. Avidians replicate by executing their code, copying instruction after instruction into fresh memory that is obtained by elongating their memory space, and then dividing off the copy.

Mutations. Digital offspring can differ genetically from their parents for several reasons. The most common cause is a single substitution error during the copying process, which is due to an inherent inaccuracy in the copy instruction.

Selection. This can be viewed as a filter, a kind of semipermeable membrane that lets some information flow into the genome, but prevents it from flowing out.

In a typical application on the evolution of "digital organisms", a population on the order $10^3 - 10^4$ programs is involved, each of them corresponding to the genome of a single avidian. The evolution of the system, followed at different intervals of time, proceeds as illustrated in the example in Fig. 7.1, which simulates the

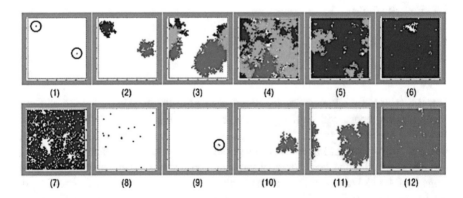

Fig. 7.1 Snapshots of an Avida population evolution in a 60 × 60 grid. The system's evolution demonstrates the presence of the largest sensed value (blue) and the next largest value (red). (*Philip McKinley, Betty H.C. Cheng, Charles Ofria, David Knoester, Benjamin Beckmann, and Heather Goldsby*, IEEE Computer Society, 2008 IEEE)

behaviour experimentally observed in a Petri disk, which is a shallow, cylindrical, glass-lidded dish that biologists use to cultivate cells. As time passes, the organisms replicate, while mutations produce variations within the population. Over generations, the presence of natural selection can generate complex behaviours, sometimes revealing unexpected peculiarities concerning the alternation of growth and stagnation.

Of course, by analysing the obtained results, we cannot neglect their compatibility with thermodynamics, which, due to the tyranny of the second law, imposes constraints on natural processes. *"Call it entropy"*, said von Neumann to Shannon when they observed the mathematical expression (2.8) adopted to measure the information. In fact, its tight connection with the equation introduced by Gibbs for evaluating entropy was evident, as illustrated in Chap. 2. The way to calculate the entropy of the outcomes of the numerical simulations conducted on an ensemble of avidians is given in Box 7.1. Starting from an initial random distribution of the bases in the genome, a high value of information is obtained, which identifies with the entropy of the system according to the previous definition. As the simulation progresses over time, the value of the entropy decreases. At first sight, the drop in entropy appears enigmatic, because it stands in contrast to the second law of thermodynamics. Actually, within the frame of Darwinian selection, such a decrease is a consequence of the introduction of a mechanism that only allows the occurrence of transitions that increase the ability of an organism to be preserved. So, it saves the information that would be lost in a mutation that corrupts the fitness. In other words, the action of natural selection can be compared to the action of a Maxwell Devil, as illustrated in Fig. 7.2. In it, the information transmitted from outside that concerns the adaption to the environment is associated with a decrease in entropy. In other words, the Darwinian selection acts as a filter, because only the mutations that reduce the entropy are kept, while the others are purged. The resurrection of the Maxwell expedient leads back to the first chapter, in which the presence, and the action, of a

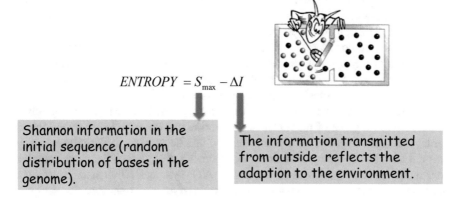

$$ENTROPY = S_{max} - \Delta I$$

Shannon information in the initial sequence (random distribution of bases in the genome).

The information transmitted from outside reflects the adaption to the environment.

Fig. 7.2 Selection favours the survival and reproduction of organisms that are best adapted to their environment. Darwinian selection acts as a filter, because only those mutations that reduce entropy are kept, while the others are purged. Thus, it behaves like a Maxwell Devil

devil was justified without violating the second law, by appealing the intervention of a ratchet mechanism similar to the one present in watches, which allows continuous motion in only one direction while preventing motion in the opposite direction, as happens to kinesin, according the analysis performed in the first chapter. Is it then possible to invoke evolutionary ratchets? In more general terms, is it possible to impose directionality on neutral processes, because the changes in DNA sequence are neither beneficial nor detrimental, to the ability of an organism to survive and reproduce? Maynard Smith and Eors Szathmary, both well-known for their contributions to theoretical biology, in a book entitled *"The Major Transitions in Evolution"* published in 1998, proposed a mechanism, called the contingent nature of irreversibility, for pushing life toward higher levels of organization. In it, it is assumed that if an entity has replicated for accidental reasons as part of a larger whole for a long time, it may have lost the capacity for independent replication. An interesting point of view was introduced about 20 years ago by Arlin Stoltzfus (see Lukes Julius and other), through an approach called Constructive Neutral Evolution (CNE), which combines epistasis and selection by accounting for the fact that cellular functions will inevitably depend on the interactions of more components. This approach has been the object of debate, because it challenges the idea that the evolution of complexity is only adaptive, selection being the sole source of creativity, the only force that can cause trends or build complex features. In other words, the proposal departs from the evolutionary conception underlying neo-Darwinism. The new proposal instead suggests that neutral evolution may follow a stepwise path to extravagance, because mutation is not a source of raw materials, but rather an agent that introduces novelty, while selection is not an agent that shapes features, but rather a stochastic sieve. Therefore, the system behaves as a unidirectional evolutionary ratchet leading to complexity, if complexity is identified with the number of components or steps required to carry out a cellular process.

The previous approach can offer a way to deal with the so-called irreducible complexity used to describe the characteristic of particularly complicated systems, whereby they need all of their individual components in just the right place in order to function, similarly to the Rube Goldberg machines mentioned in Chap. 6 in the analysis of the Z-scheme of photosynthesis.

The real question is the degree to which cellular complexity is the result of adaptation through natural selection so that the only unknowns are the identities of the selective agents, as opposed to the presence of a stochastic noise linked to the multiple interactions associated with the intrinsic complexity of the macromolecular environment. Even more challenging is the need to determine whether the involvement of neutral paths of the molecular evolution can be extended to even higher-order biological features, such as intracellular architecture, particularly by asking if natural selection is a sufficient explanation for the evolution of the more complex molecular machines, such as ribosomes or ATP synthase. The existence of such complex cellular features do not imply that each of the many changes that sculpted their structures over evolutionary time was adaptive at the time of their establishment. Such uncertainties remain a major challenge for evolutionary cell biology.

7.5   Meta-Biological Darwinism

David Hilbert, recognized as one of the most influential mathematicians of the nineteenth and early twentieth centuries, declared more than a century ago the presence of a theory of everything in the arcane universe of mathematics. This happened at the Second International Congress of Mathematicians in 1900 in Paris, where he posed 23 problems that he felt were of the highest importance. The tenth was a challenge to demonstrate the consistency of an axiomatic approach to mathematics, by reducing it to pure logic. The implied project was to formulate a finite set of axioms from which it could be possible to deduce all mathematical truths by tediously following the rules of logic.

Ironically, such a proposal led to exactly the opposite.

One motivation for developing a similar axiomatic system is to determine precisely which properties of certain objects can be deduced from other properties, called axioms. The goal then is to choose a certain fundamental set of axioms from which the other properties can be deduced. An axiomatic system is complete if every true statement can be proven from its axioms. Thirty years later, in 1931, the 25-year-old Kurt Gödel, now considered one of the most important logicians to have lived up to now, proved that for any self-consistent axiomatic system, such as the one proposed by Hilbert, there are true propositions that cannot be demonstrated. In other words, there are mathematical statements whose validity cannot be decided without using methods from outside the logical system in question. In very simplified terms, the situation is similar to that of someone who wants to demonstrate the truth of the following sentence: "What I say is not true!" Taking inspiration from this result, Gregory Chaitin, an American-Argentine mathematician, suggested that what Gödel discovered was just the tip of the iceberg, because an infinite number of true mathematical theorems exist that cannot be proved from any finite system of axioms. So, he stressed that mathematicians, instead of attempting to prove everything within the framework introduced by Hilbert, should, at least sometimes, add new axioms to the present list of official mathematics. Customarily, mathematicians are reluctant to do that, in contrast with physicists, who are pragmatically used to taking advantage of plausible reasoning instead of rigorous proof, by adding new principles to their discipline and then opening new perspectives. This is a lot of work to do! But why don't we start with biology? Biology is the kingdom of complexity, for its absence of universal rules and abundance of exceptions. Therefore, it is open to any kind of mathematics able to offer a valuable solution to its complex problems. More specifically, if we limit our analysis to a fixed set of possible genes, it is difficult to expect to be able to model the major transitions in biological evolution, such as the passage from single-celled to multicellular organisms. Contrastingly, if we are not interested in realistic simulations of biological systems, but instead want to understand biological creativity, the software space, capable of hosting both the required information and the algorithms involved in the possible transformations, is sufficient to do this. As illustrated in Sect. 6.4, DNA is essentially a program that contains information and is able to compute the behaviour of biological organisms and their functioning, by confirming that the information theory is close to biology. Furthermore, there are many provocative analogies between DNA and large, old pieces of software, similar to the ones involved in the afore-mentioned evolutionary bricolage. But how to model the involved transformations? Take a single organism, and perform random mutations on it until you get a fitter organism that replaces the original one, and then continue as before. The result is a random walk through software space with increasing fitness. In this framework, the metabiology is a computational theoretic approach to evolution introduced by Chaitin himself with the goal of extracting the essence of evolution, formalizing it, and providing the mathematical proof that it is working. By taking advantage of a model, it is simple enough to formulate heuristic arguments on the behaviour of the underlying systems. His opinion was: "*if Darwin's theory is as simple, fundamental and basic as its adherents believe, then there ought to be an equally fundamental mathematical theory about this that expresses these ideas with the generality, precision and degree of abstractness that we are accustomed to demand in pure mathematics.*"

The spark to undertake the work was offered by a sentence by Maynard Smith, who stated that "*a living organism is an evolving system where inheritances and mutations are present*". Chaitin recognized in this definition the point needed to formulate a model able to demonstrate the necessity of Darwinian evolution, by dealing with the random evolution of an artificial form of software, that is, a computer program, rather than natural software (DNA). In fact, DNA behaves like the language of a program that acts within cellular wetware, where it transmits to the molecules of amino acids the instructions with which the proteins are constructed. Within this framework, life is metaphorically compared to an evolving form of software, which, without reproduction, is a body that develops over time. That is:

$$\text{program } (10011\ldots) \rightarrow \text{computer} \rightarrow \text{output}$$

$$\text{DNA } (GCTATAGC\ldots..) \rightarrow \text{development} \rightarrow \text{organism}$$

In essence, it is a toy model designed to offer a mathematical shape to the theory of evolution by employing an evolving software, subject to mutations.

The process occurs through a random walk over hills, occurring as follows. Let us start with a single software organism and subject it to random mutations until a fitter organism is obtained. Then, we subject that organism to random mutations until an even fitter organism is obtained, and so on. A key element of the process is the introduction of fitness, which keeps organisms from stagnating by forcing them to evolve. If something challenging to do is offered, the result obtained from each organism can be expressed through the production of a positive integer. The simplest way to proceed, according to Chaitin, is the use of the so-called Busy Beaver problem aimed at generating the largest positive integer number, produced by a program whose size is less than or equal to N bits. The larger the integer, the higher the fittingness of the organism. In this approach, it comes out that the afore-mentioned approach can be utilized to express an unlimited amount of mathematical creativity. Three paths can be compared:

- the first occurs through a set of random choices, each of them corresponding to the mutations in the neo-Darwinian theory. All are accepted, despite the positive nature of the outcome with respect any goal of the program itself.
- the second occurs through a random walk, in which any favorable mutation is incorporated into the program, so that it is modified or evolves as an organism, consistent with the action of the selection in neo-Darwinian theory.
- the third occurs through a controlled walk, because each step is performed in accordance with an intelligent choice.

The effectiveness of the processes is determined by comparing the results of each path with the one obtained from the afore-mentioned typical reference program, called the Busy Beaver. It turns out that while the first choice represents a loss with respect to any evolutionary perspective, the second path, let us call it Darwinian, provides results that are obviously less effective than those of the third one, in which any choice is controlled to be oriented towards the best result. But the second is not that far from the third.

The interest of the work is associated with the heuristic use of a scientific metaphor that, by accounting for some experimental observations, builds an abstract model able to provide interesting information about the behavior of reality. The limit of the approach is the implicit assumption that there exists a direct correlation between functional operation and structural organization, so that the increase in the former is necessarily accompanied by an increase in the latter.

Box 7.1 Entropy of a Genomic Sequence

In a genome in which a sequence of N sites i is present, each of them can take on four bases of DNA, with probabilities $\{pC(i), pG(i), pA(i), pT(i)\}$, where C stands for cytosine, G for guanine, A for adenine, T for thymine. The entropy of a sequence whose elements are distributed in the N sites, each with the previous probabilities, is calculated by means of Eq. (2.8), so that

$$S_i = - \sum_{j}^{C,G,A,T} p_j(i) \log p_j(i). \tag{7.1}$$

The overall entropy is then given by

$$S = \sum_{i} S_i. \tag{7.2}$$

References

Chaitin, Gregory. *Proving Darwin. Making Biology Mathematic*, Pantheon Book, New York, 2012.

Adami Christoph. *The use of information theory in evolutionary biology*, Ann. N.Y. Acad. Sci. 1256 (2012) 49–65.

Adami Christoph, Charles Ofria and Travis C. Collier. *Evolution of Biological Complexity*, Proc. Nat. Acad. Sci. USA 97 (2000) 4463-4468.

Daniel Dennett. *Darwin Dangerous Idea: Evolution and the Meaning of Life*, Simon and Shuster, New York, 1995.

John E. Mayfield, *The Engine of Complexity, Evolution as Computation*, Columbia University Press, New York, 2013.

Lukes Julius, John M. Archibald, Patrick J. Keeling, W. Ford Doolittle and Michael W. Gray. *How a Neutral Evolutionary Ratchet Can Build Cellular Complexity*, Life, 63(7): 528–537, July 2011.

Lynch, Michael, Mark C. Field, Holly V. Goodson, Harmit S. Malik, *Evolutionary cell biology: Two origins, one objective,* 2014 https://doi.org/10.1073/pnas.1415861111

Ridley Matt.*The Evolution of Everything*, Harper, New York, 2015.

Dawkins Richard. *Climbing Mount Improbable*, W.W.Norton, New York, 1996.

Shanahan Timothy. *The Evolution of Darwinism*, Cambridge University Press, 2004.

Mustonen Ville, Michael Lassig. *Molecular Evolution under Fitness Fluctuations*, PRL 100, 108101 (2008).

Chapter 8
Imitation of Life: The Cell in a Silicon Chip

8.1 The Chemoton

"I thought there couldn't be anything as complicated as the universe, until I started reading about the cell", wrote the astrophysicist Eric de Silva.

Despite this daunting sentence, the creation of a cellular model through the extended application of mathematics to biology remains a challenging and stimulating task, as summarized in Fig. 8.1. The development of a computational approach to biology, customarily called "in silico", involves the computer simulation of cellular subsystems by accounting for the complex connections engaged in their functions, including:

- metabolism, composed of a set of bio-chemical reactions, by which the cell sustains its growth and energy requirements;
- signal transduction, which is the process by which a chemical or physical signal is transmitted through a cell as a series of molecular events;
- gene expression, which is the process by which genetic instructions are used to synthesize gene products, usually proteins;
- motility, which is the capability of cells to exhibit self-generated, purposeful movement;
- cell division and differentiation.

The first significant attempt at cell simulation was pursued by Tibor Gànthi, a Hungarian chemical engineer, who, in 1966, published a book entitled "*The Principle of Life*". In it, a subtle analysis of cellular behaviour is offered. In fact, a characterization of living systems was proposed, together with the introduction of a simplified mathematical model of their behaviour. Baptized a 'chemoton', short for "chemical automaton", the model can be summarized through the three coupled autocatalytic subsystems illustrated in Fig. 8.2, reflecting, respectively:

© Springer Nature Switzerland AG 2018
S. Carrà, *Stepping Stones to Synthetic Biology*, The Frontiers Collection,
https://doi.org/10.1007/978-3-319-95459-2_8

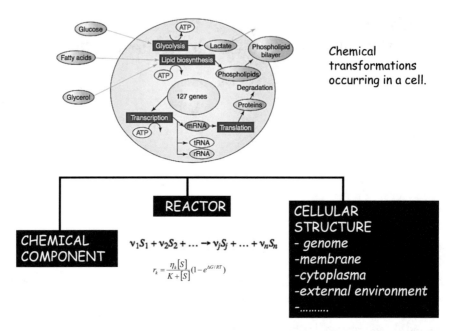

Fig. 8.1 The simulation of a cell's behaviour is a challenge for Chemical Engineering, because the cell can be compared to a reactor in which a very complex network of interconnected chemical reactions is present

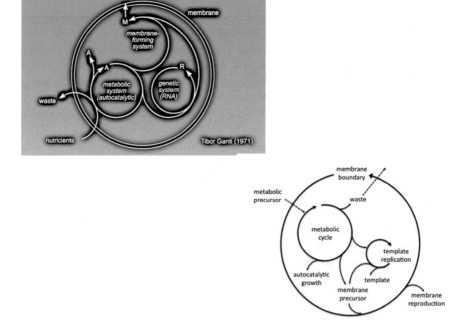

Fig. 8.2 The Chemoton model, made of three tightly coupled subsystems: (1) the autocatalytic metabolic cycle, (2) the informational cycle that generates the polymer, (3) the membrane-forming subsystems

– a self-reproducing metabolic chemical network, which transforms the external
  nutrients into the internal material required for replication and membrane growth;
– a polymerization subunit, including the self-replication cycle of a homopolymer
  that produces specific precursor molecules necessary for growth of the
  membrane;
– a membrane subsystem that encloses all of that.

Thanks to the presence of a controlling system, the monomers start to polymerize
only when a threshold value of its concentration has been reached, by producing the
molecules that are spontaneously incorporated into the membrane.

Despite its apparent simplicity, the chemoton model deserves attention, because it
represents the basic characteristics that any living system capable of evolution must
satisfy. Gánti's major interest was the description of the minimal system able to
reflect the functional properties of life on Earth, through an abstract chemical
dynamical model, which fulfils the following criteria, considered essential for life:

1. Unity;
2. Metabolism;
3. Stability;
4. Information;
5. Growth and multiplication;
6. Hereditary system enabling evolution;
7. Mortality.

The ability to generate a vast number of hereditary variations is the hallmark of all
of the known forms of life and depends on the previous criteria. In spite of its
simplicity, the chemoton model has been and it is yet considered a useful guide for
any approach to the origin of life. Also rather interesting is the fact that the chemoton
model sheds light on the emergence of functional organization. In fact, when a
system obtains the organization to effect its own long term self-maintenance, the
parts and the processes it generates acquire a function. Accordingly, the biological
function can then be identified with the role that a part, a process or a mechanism
plays within a system, by contributing to a goal-directed behavior of the system
itself.

8.2 A Challenging Problem

Masaru Tomita, a pioneer on the subject, argues that the whole-cell simulation is a
major challenge of the present century. But why simulate a cell? Because the
availability of a good model of a cell can accelerate biological and bioengineering
discoveries by facilitating experimental programs and their interpretations. Moreover, it allows us to perform iterative testing versus experimental information by
enriching our understanding of biological systems. The obtained knowledge, in
combination with the recent de novo synthesis of genomes and their transplantation

Table 8.1 Cellular processes

Process type	Dominant phenomena
Metabolism	Enzymatic reactions
Signal transduction	Molecular binding
Gene expression	Molecular binding
DNA replication	Molecular binding, polymerization
Membrane transport	Osmotic pressure, membrane potential

to produce synthetic cells, raises the exciting possibility of using models to enable computer-aided rational design of novel microorganisms. The most important cellular processes, shown in Table 8.1, are involved in the production, interconversion, transport, or consumption of the species within the cellular network. Therefore, their occurrence incorporates a wide range of interactions, which includes the formation or breaking up of chemical bonds, the occurrence of catalytic reactions, and the regulation of biomolecular activities.

The reliability of the models employed to simulate such processes has been strongly improved, because a vast amount of basic genetic and biochemical information is becoming rapidly available, thanks to the contribution of the omics techniques mentioned in Sect. 4.7. For simple systems, the resulting behaviour can be intuitively understood, but for more complex networks, especially those involving feedback, an accurate description of system behaviour is only possible by accounting for the description of the various interactions in quantitative terms by means of mathematical models, exactly as happens in other fields of science and engineering that are making wide use of mathematical simulations. For instance, chemicals are obtained from highly integrated industrial processes with complex structures that rival those of living cells. In fact, the modern chemical processes include many interconnected reactions, which can be addressed only through detailed, and often sophisticated, mathematical computation. Similarly, any model of cellular phenomena takes the form of schematic interaction diagrams whose components are molecular species, such as ions, small molecules, macromolecules, or molecular complexes.

Cells are probably the most complete example of traffic of signals, comprising millions of molecules that act coherently, even if the system is far from equilibrium, through the exchange of matter, energy and information with the environment. All of these molecular processes, ultimately controlled by genes, take place at different points in space and time and involve the leading participation of proteins, which also act as nanomachines, as occurs, for instance, with the kinesins and myosin that drive the cellular transportation processes. The cellular network can be divided into three major self-regulated subnetworks:

- **genome**, in which each gene can express the process by which information is used in the synthesis of functional products as proteins;
- **proteome**, defined by the set of proteins, their interactions and their catalytic activity;
- **metabolome**, which includes the network integrating all of the metabolites and the pathways that link each other.

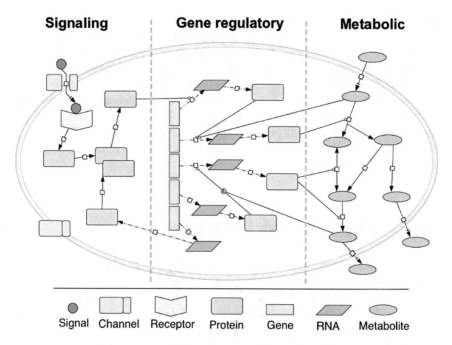

Fig. 8.3 The main processes occurring inside a cell. Signals from the environment propagate through a cascade, leading to gene activation. The gene regulatory networks control the transcription level of genes, which are transcribed into RNA molecules, and subsequently translated into proteins involved in all cellular functions. (*Machado et al. AMB Express 2011, 1:45*)

All of them, as illustrated in Fig. 8.3, are broadly intertwined, because genes can affect some metabolic pathways, while proteins catalyze the formation of nucleotides, in turn affecting the processes of translation. Particularly important is the presence in the cells of the gene regulatory networks that control the transcription of genes into RNA molecules, subsequently translated into the proteins involved in all cellular functions. At any rate, the main cellular processes that occur inside the cell imply the intervention of signals coming from the environment, which propagate through a signaling cascade, leading to gene activations, as summarized in the aforementioned figure.

Although all of the previous processes contribute to the development of the "in silico" biology, at present, it is difficult to merge different subsystem models into one single-cell model, so it is appropriate to start with a more simple approach focused on specific subsystems. Let us begin with a simulation of the behaviour of a self-surviving cell. Of course, a constant supply of energy through ATP is required to maintain protein and membrane synthesis, so that glucose is essential for the survival of the cells. The model must then account for the uptake of glucose into the cytoplasm, where it is metabolized through the glycolysis pathway with the production, as illustrated in Sect. 4.6, of ATP, which is the energy source mainly employed for the synthesis of proteins, because the cell must constantly produce them to sustain life. The genes are transcribed by m RNA and then translated into proteins by ribosomes. The membrane structure of the cell must also be modeled by accounting for its degradation over time. Thus, the presence of a pathway for the required biosynthesis of the cell membrane is occurring through the formation of the fatty acids and glycerol that generate the phospholipid bilayers. The afore-mentioned processes, already described in the previous chapters, are summarized in the same figure, through demonstration of their connections. The formulation of an appropriate mathematical model then proceeds through the compilation of the equations expressing the balance of the various components involved, which can be pursued at different levels of accuracy.

8.3 Modeling of Metabolic Reaction Systems

The cell's metabolism, which occurs through a network of chemical reactions catalyzed by enzymes, can be compared to a chemical engine that converts raw materials into energy and is, in molecules, required to build the biological structures. In a metabolic network, the substrates enter the cell and are converted into products.

But what is happening inside? A set of intracellular reactions are taking place, each of them with a specific rate r_k, customarily expressed by means of the Michaelis–Menten Eq. (4.2):

$$r_k = \frac{\eta_k [S]}{K + [S]} \left(1 - e^{\Delta G/RT} \right). \tag{8.1}$$

The term inside the brackets on the right hand side is due to the presence of the inverse reaction. The free energy change of the reaction ΔG, which depends on the composition according to Eq. (4.2), is equal to zero at the thermodynamic equilibrium point, so that the influence of the k-th reaction disappears.

The description of the evolution of the system can be borrowed from the chemical reaction engineering approach, in which a cell, in an approximate approach, is compared to a stirred, or well-mixed, micro-reactor in which the concentrations of *the components are assumed to be uniform*. It follows that the model equations, expressing the material balances of the various components, are formulated by applying the mass conservation equation introduced in Chap. 2:

$$\frac{dC_i(t)}{dt} = \sum_{ex} \Phi_{ex} C_i^{ex} + \sum_k \nu_{ki} r_k \tag{8.2}$$

time variation = exchanges + chemical reactions.

The previous expression is similar to Eq. (2.7), and in it, ν_{ki} is the stoichiometric coefficient of component i in the k-th reaction. A system of non-linear ordinary differential equations (ODEs) is obtained able to describe the variation of the amount

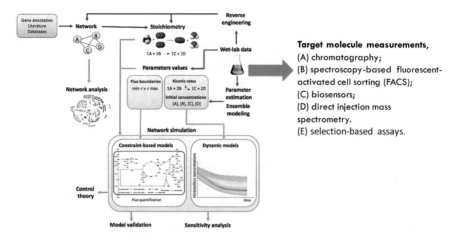

Fig. 8.4 The approach and strategy able to collect the experimental information required to simulate a bio-reactive system. (*P. Cazzaniga et al., Metabolites 2014, 4, 1034–1087*)

of each species in the modeled system as a function of time. This approach can be applied to all kinds of biological pathway for modeling dynamical systems in several areas. If the kinetic models are available, the time-course simulations can predict the response to different inputs and design system controllers. However, the building of the model requires insight into the reaction mechanisms so as to be able to select the appropriate rate laws. Therefore, the most demanding part is the obtainment of the experimental data required for the evaluation of the parameters of the reaction rate equations. A clear and effective illustration of the approach and strategy that can be followed to collect the information required to simulate the system has been suggested by *P. Cazzaniga et al.* This is illustrated in Fig. 8.4, where the different colors indicate the paths to be pursued. Particularly:

- *Violet arrows* are related to the reconstruction of the network of the different reactions by relying on pre-existing knowledge and on the new available experimental data about cellular processes, as well as by exploiting various databases that contain information on molecular interactions and pathways.
- *Red arrows* are involved in the use of experimental data to estimate the system parameters.
- *Green arrows* concern the computational analyses of metabolic networks and models,
- *Blue arrows* concern the computational simulation of different types of model.

The formulation of the model, illustrated in Fig. 8.5, proceeds throughout the following steps:

- **Topology**. A reaction network able to describe the pathway of the entire reacting system is explored and then fixed. In other words, by stating the way in which the

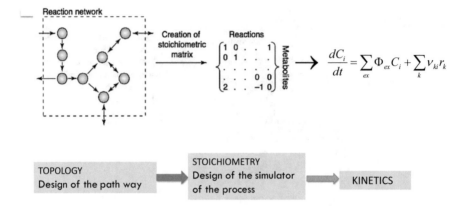

Fig. 8.5 The steps involved in elaborating the kinetic model of a cell. This includes the different capabilities required to approach the problem, concerning, respectively, the topology of the reaction network, the stoichiometry of the involved reactions and their rate laws

chemical transformations of the different compounds are occurring and what the possible interconnections between them are.

- **Stoichiometry**. This is aimed at designing the simulator of the process by creating a non-symmetric matrix **S**, called the stoichiometric matrix, in which each row corresponds to one of the metabolites involved in the network, while in the columns, the stoichiometric coefficients of the metabolites in the different reactions are given. Of course, there is a negative coefficient for every metabolite consumed, and a positive coefficient for every metabolite produced. A stoichiometric coefficient equal to zero is used for every metabolite that does not participate in a particular reaction. Thus, **S** is a sparse matrix, since most biochemical reactions involve only a few different metabolites. **S** contains all of the information about how the substances are connected through the different reactions present in the network, and thus it characterizes the topological structure and architecture of the network.
- **Kinetics**. The expressions of the rates of the involved reactions are ordered vertically in a column vector.

The consequent strategy for the formulation of the model equations, as illustrated with a simple example in the figure, leads to a system of ordinary differential equations, each expressing the material balance of a given component. Its integration, which can be easily pursued through a numerical approach, provides the evolution over time of the various components. Moreover, it is important to stress that the cells are subject to both invariant and adjustable constraints. The former are physico-chemical in origin and include the stoichiometric and thermodynamic constraints. Adjustable constraints will instead change in a condition-dependent manner, due to the action of the Gene Network Regulator (GNR), which is collections of genes combined together to form networks that provide instructions for making proteins, consistent with the scheme illustrated in Fig. 8.3. In a regulatory operation, for instance, a gene G is transcribed to produce an enzyme E, which catalyses the reaction, converting the substrate A into the product B, which, in turn, affects the transcription of G, leading to depletion of E. These regulatory proteins interact with those of the metabolic system, acting as enzymes, by modifying their activity. Acting as on–off switches, genes can modify the metabolic paths by promoting or suppressing the activities of proteins themselves, determining the behaviour of each cell in a way similar to an electronic circuit, similarly to the behaviour of the operon, as described in Sect. 5.7. In other words, the GNR provides the instructions required to make proteins, able to modify the metabolic paths by promoting or suppressing the activities of the existent proteinic enzymes.

As an example of application of the introduced concepts and procedures, let us recall the photosynthetic process. The overall process has been illustrated in Chap. 6, but let us now focus our attention on the dark phase, in which the different reactions are organized in the metabolic cycle proposed by Calvin and Benson, shown in Fig. 6.6. It is a central part of the carbon metabolism that occurs in the vegetable world, involving 11 different enzymes that catalyze 13 reactions and occurring in three phases. The capture and fixation of carbon dioxide takes place thanks to the contribution of RuBisCo, which is the most abundant catalyst present on the earth. The afore-mentioned three main phases and outputs of the Calvin–Benson cycle are common to cyanobacteria, the phylum of bacteria often called blue–green algae, which obtain their energy through photosynthesis by using a specific type of chlorophyll, together with other pigments able to acquire energy from solar light. The initial phase is the fixation of CO_2 into a carbon skeleton, the second phase describes the reduction of the 3-phosphoglycerate (PGA), while the final regeneration phase of the cycle involves several reactions, leading to the reassembling of the Ribulose bisphosphate, as shown in the afore-mentioned figure. The overall photosynthetic process is driven by the flux of solar energy, subject to a periodic variation due to the passage from day to night. The mathematical model of the system is therefore expressed by a system of ordinary differential equations obtained by applying Eq. 8.2 to each of the components present in the cycle. Their non-linearity is a consequence of the non-linearity of the rate law of each of the involved reaction S. Its integration reveals the presence of oscillations versus time of the chemical species concentrations, by introducing a typical characteristic of the complex dynamical systems, as expected for a process as periodic as photosynthesis. In agreement with the behaviour of natural systems, leaves show oscillations of CO_2 uptake, of O_2 evolution, and of chlorophyll fluorescence when light varies. Due to its presence in the core part of a variety of models of the entire photosynthetic processes, the Calvin–Benson cycle is the best-studied metabolic system of the vegetable world. Actually, anyone who intends to model such a system is compelled to look at existing models so as to adapt, refine and improve them.

8.4   FBA Approach

The introduced method starts from an approach based on accurate knowledge of the rate equations. However, their availability, as shown, requires an evaluation of the corresponding kinetic parameters through detailed experiments, such that an alternative simplified and approximate approach, called Flux Balance Analysis (FBA), has been introduced. It was applied to the prediction of the growth of bacteria, particularly *Escherichia coli*, but has subsequently been extended to many other contexts for the purpose of analyzing the capabilities of organisms subject to different environmental and genetic perturbations.

The first step in FBA is, again, the mathematical representation of the metabolic reaction systems in the form of the introduced stoichiometric matrix **S**, which imposes the constraints on the flow of metabolites in the frame of the network. The FBA deals with the solution of the problem by assuming that the metabolic network is subject to three conditions:

- The network is at steady–state.
- The reaction rates are expressed in the linear approximation through simple equations of the kind $r = kC$.
- The enzyme catalytic activity in each reaction is limited to an admissible range by fixing the conversions in order to respect the presence of an equilibrium point in the chemical reactions.

Their application, as a consequence of the previous constraints, results in a system of n linear algebraic equations with respect to the m values of the concentrations of the considered compounds. In any realistic large-scale metabolic model, there are more reactions than compounds. In other words, the number of equations is higher than that of unknown variables, so that there is no unique solution to the system of equations. Actually, the interest is mainly focused on the identification of the maximum growth rate of an organism or its maximum ATP production, within the frame of a particular set of constraints.

Thus, the constraints confine the steady–state fluxes to a feasible set, however, they do not yield a unique solution. The determination of the metabolic flux distribution can then be pursued only through the solution of an optimization problem by seeking the maximum of an advisable objective function, under given constraints. The underlying idea of FMB is that if a reaction network does not have a unique solution, it may well have a best solution. The approach is to select an optimized linear combination of metabolic fluxes, consistent with the experimental cellular growth.

Substantially, the process for finding the solution is much like the algorithm used to optimize the operations performed on a chemical plant or on oil refineries. Suppose a refinery has a range of products (gasoline, diesel fuel…) that differ in manufacturing cost and market value. The mathematical technique of linear programming computes a mix of products that maximizes profit. Applying the same method to the living cell, it yields a set of reaction rates that make the most efficient use of available resources, such as nutrients. Also, if it is not known whether the micro-organisms optimize their growth in this way, it appears to be a plausible assumption in the context of Darwinian evolution.

Given the above, within the frame of the FBA model, the regulatory operations can be described by making use of Boolean algebra, that is, by restricting the expression of a transcription unit to the value 1 if it is transcribed and 0 if it is not. In this extension, called rFBA (r for regulatory), the transcription factors can be active or inactive in the regulation of the enzymes that catalyse the metabolic reactions. In other words, the production rate of a metabolite can drop to zero if the enzyme that produces it is not transcribed.

8.5   Exploring the *Escherichia coli*

We already encountered *Escherichia coli* as one of the most important model organisms in biology. In fact, researches into its metabolic structure and its behaviour aided in the development of microbial system biology. The history of the reconstruction process of the metabolic network of *Escherichia coli* spans more than a decade. Just to mention a few steps from the research, it should be remembered that Markus W. Covert and Bernhard Palsson used rFBA to model the regulation of the central metabolic network of *E. coli*, which includes 149 genes, 16 regulatory proteins, 73 enzymes, 45 transcriptional regulations and 113 biochemical reactions. The growth predictions agreed well with many experimental measurements. A more comprehensive model that accounts for 1010 genes was later introduced, also by Covert along with some coworkers. Reconstruction efforts were built off of successive versions, each adding new subsystems, particularly the cell wall synthesis. On the whole, the formulation of the genome scale model of *Escherichia coli* was focused on six of its applications:

1. **metabolic engineering**. The native biochemical pathways of bacteria can be manipulated to produce industrial and therapeutically relevant compounds in efficient ways. This topic will be the subject of the following chapter.
2. **model-driven discovery**. Some aspects of bacterial functions remain uncharacterized. In order to expand the current understanding on the subject, an iterative approach can be pursued that, by taking as a reference the genome of *Escherichia coli*, encourages researchers to decrease the gaps in knowledge on other bacteria.
3. **prediction of cellular phenotypes**. The phenotype of a cell includes the observable characteristics resulting from the interaction of its genotype with the environment. The diverse set of biochemical pathways in bacteria have conferred a vast phenotypic potential that has enabled them to thrive in different environments, ranging from volcanic vents on the bottom of the ocean to clouds, glaciers, and the human gut. Constraint-based modeling that takes advantage of the progress in the genome of *Escherichia coli* allows for the rapid prediction of growth of phenotypes under various genetic and environmental conditions.

4. **analysis of biological network properties**. In order to elucidate the relationship between the network structure and function, researchers have turned to network analysis in which biochemical reactions are transformed and explored using mathematical methods to arrive at biologically insightful conclusions.
5. **studies of evolutionary processes**. This includes exploration of losses-of-function and mutations through the addition of new reactions by horizontal gene transfer and gene duplication. *Escherichia coli* has been proven to be useful in modeling microbial evolution through the elimination and addition of new metabolic network content.
6. **models of interspecies interactions**. This provides a platform that would allow for the prediction and simulation of biological interactions. It fulfills the interest in a better understanding of the host–pathogen interactions for the development of improved antimicrobials and the use of microbes for environmental remediation. Such categories are summarized and illustrated in Fig. 8.6.

8.6   Whole Cell

An important step forward in the simulation of life was achieved in 2013, when a paper of the results of a research conducted by Markus Covert, together with eight coworkers, was published. In it, the characteristics and performances of a computer program, called "Whole Cell simulation", were communicated, together with its application towards a description of the full life cycle of a prokaryotic bacterium. The progress with respect to the previously described approaches is relevant, because the model includes all of the major processes of bacterial life, including the transcription of DNA into RNA and its translation into protein, the metabolism of nutrients to produce energy and structural constituents, the replication of the genome and, finally, the cell's reproduction through its fission. The choice of the bacterium *Mycoplasma genitalium* as a protagonist of the research was reminiscent of a proposal advanced by Harold Morowitz, an American biophysicist well-known for his researches into the applications of thermodynamics to living systems. Such a choice turned out to be attractive and effective, because it was made on the smallest existing self-replicating organism. Its diameter, roughly half a micrometer, is one fourth of the diameter of the familiar *Escherichia coli*, which is two micrometers long. First observed in 1980, it is now classified among the bacteria with a lipid membrane, and is thus highly resistant to antibiotics. As a rule, the formulation of a computer model implies a multitude of choices and compromises concerning the appropriate level of details required to describe the significant events, molecular events in the present case. Let us, for instance, recall the oxidation processes that take place in living organisms through the transformation of ingested glucose. As discussed in Chap. 4, the number of involved components is elevated, and each of them turns into each other in a cyclic succession. The final products are carbon dioxide and water, so that the global reaction can simply be described by the

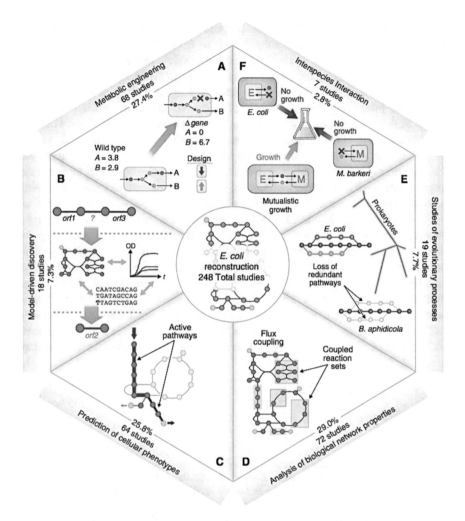

Fig. 8.6 Trends to explore in regard to the interaction of the *Escherichia coli* metabolic network with other organisms under different environmental conditions. *(Douglas McCloskey Molecular Systems Biology 9; Article number 661; doi:10.1038/msb.2013.18)*

comprehensive combustion reaction, which doesn't reveal much about what is actually happening inside of the cell. Actually, a closer look must account for the several intermediate steps present in the Krebs cycle, as shown in Fig. 4.2.

The "Whole cell" model has been formulated with different levels of detail, because certain key macromolecules are considered identifiable entities, while smaller molecules are lumped together. For instance, ribosomes are individually represented, each with its own history, while no representation is given to the amino acids that are linked together to form a protein. The model is structured on a set of modules, each corresponding to major cellular activities, such as replication of the genome and the synthesis of proteins. Moreover, it is based on experimental data collected from almost one thousand publications from which numerical values were extracted to be employed as parameters of the model. One module is devoted to metabolism, in which most of the classical biochemistry happens. The cell is compared to a well-stirred chemical reactor, where all molecules have the same chances of interacting, regardless of their location. Their balance is pursued by means of the rFBA approach by including 104 enzymes, 585 substrates, 441 chemical reactions and 204 transport processes. Contrastingly, in the transcription module, where the enzymes occupy specific positions along the bacterial chromosome, the discrete events that occur are governed by probabilities evaluated to match the experimentally-observed distribution. So, the transitions between the states being random events, the runs produce somewhat different results, even with the same initial conditions and interactions with the environment, consistent with the fluctuations present in real biology. The model has been validated for a wide range of independent conditions involving its many biological functions, previously mentioned. The simulations give plausible results if the rate of growth in biomass and the concentrations of various metabolites are considered. In conclusion, the approach fulfills the yearning to understand how complex cells arise from individual molecules and their interactions. Moreover, the attempts to build a computer model of a cell are a declaration that life is comprehensible, because there is nothing supernatural in its behaviour, seeing as it can be reduced to an algorithm of a computational process.

References

Covert, Marcus W., Christophe H. Schilling, Bernhard Palsson. *Regulation of Gene Expression in Flux Balance Models of Metabolism,* J. theor. Biol. (2001) 213, 73–88.

Gánti, T., 1971, 1987. *The Principle of Life.* Omikk, Budapest, Hungary (Translated into English 1987).

Macklin, Derek N, Nicholas A Ruggero and Markus W Covert. *The future of whole-cell modeling,* Current Opinion in Biotechnology 2014, 28:111–115.

McCloskey, Douglas, Bernhard Ø Palsson and Adam M Feist. *Basic and applied uses of genome-scale metabolic network reconstructions of Escherichia coli, Molecular Systems,* Biology (2013) 9:661. www.molecularsystemsbiology.com

Simpson Michael, Gary S. Sayler, James T. Fleming, Bruce Applegate. *Whole cell biocomputing,* TRENDS in Biotechnology Vol.19 No.8 August 2001 317.

Schilling, Christophe H., Stefan Schuster, Bernhard O. Palsson, and Reinhart Heinrich. *Metabolic Pathway Analysis: Basic Concepts and Scientific Applications in the Post-genomic,* Biotechnol. Prog. 1999, 15, 296–303.

Simpson, Michael L. et al. *Engineering in the Biological Substrate: Information Processing in Genetic Circuits,* PROCEEDINGS OF THE IEEE, VOL. 92, NO. 5, MAY 2004.

Cheemeng Tan, Hao Song, Jarad Niemi and Lingchong You. *A synthetic biology challenge: making cells compute,* Mol. BioSyst., 2007, 3, 343–353.

Thiele, Ines, Neema Jamshidi, Ronan M. T. Fleming, Bernhard Ø. Palsson. *Genome-Scale Reconstruction of Escherichia coli's Transcriptional and Translational Machinery: A Knowledge*

Base, Its Mathematical Formulation, and Its Functional Characterization, PLoS Computational Biology, March 2009 | Volume 5 | Issue 3 | pag.1.

Masaru Tomita. *Whole-cell simulation: a grand challenge of the 21st century,* TRENDS in Biotechnology Vol.19 No.6 June 2001 205.

Masaru Tomita, Kenta Hashimoto, Kouichi Takahashi, Thomas Simon Shimizu, Yuri Matsuzaki, Fumihiko Miyoshi, Kanako Saito, Sakura Tanida, Katsuyuki Yugi, J. Craig Venter and Clyde A. Hutchison. *E-CELL: software environment for whole-cell simulation,* Bioinformatics, Vol. 15 no. 1 1999 Pages 72–84.

Chapter 9
Synthetic Biology at Work

9.1 The Synthetic Approach to Biology

The expression "synthetic biology" appeared for the first time in a couple of publications by the French biologist Stephane Leduc, respectively *"Théorie physico-chimique de la vie et générations spontanées"*, published in 1910, and *"La Biologie Synthétique"*, published 2 years later. Leduc attributed the afore-mentioned expression to researches into the synthesis of life from inanimate matter, by emphatically asking why the synthesis of a cell is considered less admissible than that of a molecule. Such statements did not find interest and approval among his contemporary colleagues.

The present understanding of synthetic biology emerged more than 50 years later, in *1973*, in a speech by the Polish geneticist Waclaw Szyblalski during a panel discussion at the Annual Biological Conference on Strategies for the Control of Gene Expression. Time became ripe to reiterate the term, by giving concreteness to its content, and this occurred in 1978, when Werner Arber, Daniel Nathans and Hamilton Smith won the Nobel prize *"for the discovery of restriction enzymes and their application to problems of molecular genetics"*.

On such an occasion, Szybalski wrote the following editorial comment in the journal "Gene": *The work on restriction nucleases not only permits us easily to construct recombinant DNA molecules and to analyze individual genes, but also has led us into the new era of synthetic biology where not only are existing genes described and analyzed, but new gene arrangements can also be constructed and evaluated.*

After some time, certain technical issues concerning the control of metabolic networks appeared, bringing about a significant advance in the subject. This occurred at the beginning of this century, when details of the creation of synthetic biological circuit devices, such as the toggle switch and the repressilator, were published. As a consequence of this emerging approach, engineers started to consider biology as a *technology*, within the frame of which the wide perspectives of

© Springer Nature Switzerland AG 2018
S. Carrà, *Stepping Stones to Synthetic Biology*, The Frontiers Collection,
https://doi.org/10.1007/978-3-319-95459-2_9

manufacturing materials and structures, producing energy, providing food and enhancing human health, are present, including the stimulating perspective of designing and building engineered biological systems able to process information.

The shared task was the development of synthetic entities at a high level of complexity by manipulating simpler parts that came from the preceding level, with the awareness that humans excel in the use of tools that they find by looking to nature for the required instruments. In the present case, attention was focused on the fundamental molecules of the biological world, assumed to be the bricks of the afore-mentioned new chemical engineering.

9.2   Progress in Gene Manipulations

As mentioned, the launch of the amazing era of genetic engineering coincides with the isolation of the set of proteins known as restriction enzymes, which have the peculiarity of acting as DNA scissors. They do not sniff a code randomly, but rather recognize a specific nucleotide sequence whose DNA is cut only at a specific site, which is known as the restriction site. In other words, they recognize precise letters in the entire genetic row. Such restriction enzymes, called endonucleases, are proteins produced by bacteria able to cleave DNA at a given site, in correspondence with an internal phosphodiester bond. In bacteria, they cleave foreign DNA, thus eliminating infecting organisms, so that they can be isolated from bacterial cells and used to manipulate fragments of DNA contained in genes. Actually, there is an enormous range of restriction enzymes with some cutting capacity in common sequences, but the present focus is on the ones that have the peculiarity of operating on particular pieces of DNA. As a metaphor, we can imagine an editing working on this book, which, being able to recognize the word "enzyme", would break it into fragments. Contrastingly, if the tool is able to recognize only the word "chimera", it will chop the text into no more than a couple of parts. When a restriction enzyme cleaves a restriction site, the reaction creates highly reactive "sticky ends" on the broken DNA. In other words, restriction enzymes make staggered cuts at their recognition sites, producing ends with a single-stranded overhang that, if joined together, can give rise to new DNA molecules. But how can pieces of DNA from different sources recover the structure of a single DNA molecule? The common process is based on the intervention of an enzyme called DNA ligase, which seals the gap between the molecules, forming a single piece of DNA. Its job is to join together fragments of newly synthesized DNA to form a seamless strand, so that if two pieces of DNA have matching ends, DNA ligase can join them together to make an unbroken molecule. It does this by catalyzing the reaction in which a phosphate group links with a hydroxyl group by producing an intact sugar-phosphate backbone. The process occurs thanks to the contribution of ATP as an energy source. Metaphorically speaking again, it looks like a natural editing tool that enables biotechnologists to treat DNA and genomes much in the same way we use software processing to cut, copy and paste text from one section to another. The amazing thing is that hundreds of restriction enzymes are at our disposal at low cost, able to undertake the dizziest transformations that can occur in nature.

The DNA molecules formed by laboratory methods to bring together genetic material from multiple sources, by creating sequences that would not otherwise be found in the genome, are called recombinant DNA (rDNA). Such molecules are also sometimes called chimeric DNA, being made from materials that come from two different species, similar to the mythical chimera, which, according to Greek mythology, was a fire-breathing female monster with a lion's head, a goat's body and a serpent's tail. The possibility is offered from the fact that DNA molecules present in all organisms share the same chemical structure, differing only in the sequence of the nucleotides. Emphasis must then be given to the fact that the rDNA is a piece of DNA opportunely created.

The DNA sequences used in the construction of rDNA molecules can originate from any species, for instance, plant DNA may be joined to bacterial DNA. Not only this, but sequences that do not occur anywhere in nature may be created through the chemical synthesis of DNA subsequently incorporated into recombinant molecules. In conclusion, thanks to the use of recombinant DNA technology, any DNA sequence may be created and introduced into any of the ones present in nature, opening a wide spectrum of possibility for the synthetic approach to biology.

One more step is the procedure called cloning, a term introduced by Haldane, who took inspiration from an ancient Greek word referring to the process whereby a new plant can be created from a twig. In the present case, attention is focused on molecular cloning, which refers to the process by which recombinant DNA molecules are produced and transferred into a host organism, where they are replicated. Generally, DNA sequences from two different organisms are used: the species that is the source of the DNA to be cloned and the species that will serve as the living host for the replication of the recombinant DNA (Fig. 9.1).

Most commonly, small circular, double-stranded DNA molecules, called plasmids, are present in bacteria, physically separated from the chromosomal DNA and capable of replicating independently. In nature, plasmids often carry genes, and their chromosomes are big and contain all of the essential genetic information for living under normal conditions. Plasmids are usually very small and contain only the additional genes that may be useful to the organism in certain situations. Quite interesting is the fact that artificial plasmids are widely used as vectors in molecular cloning, because they serve to drive the replication of recombinant DNA sequences within host organisms as shown in the figure.

Synthetic biology is thus taking advantage of the employment of plasmids to create new genomes able to incorporate some elements of reputed interest and then test the soundness of the predictions. Therefore, the creation of novel plasmids is being investigated for their many potential applications in the transmission of information on the modifications of existing genomes and for the purpose of opening new perspectives in synthetic biology. "Biology is technology", writes Robert Carlson to enphasize the peculiar nature of the researches in progress, whose applications are very well described in the cited book of George Church.

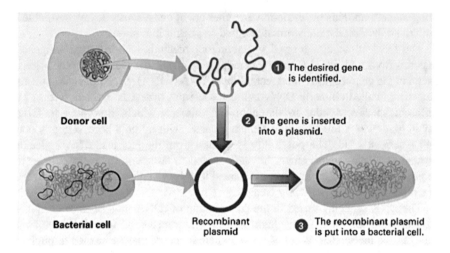

Fig. 9.1 A plasmide is a small DNA molecule within a cell that can replicate. Artificial plasmids are used as vectors in molecular cloning, because they serve to drive the replication of recombinant DNA sequences within host organisms

9.3   Frontier of Genome Engineering: The CRISPR/Cas System

The ability to make highly specific changes in the DNA sequence of a living organism by customizing its genetic makeup is defined as gene editing. It is performed using the afore-mentioned endonucleases, which deserve particular attention because they are capable of cleaving well-defined phosphodiester bonds between monomers present in nucleic acids. Nucleases variously affect single- and double-stranded breaks in their target molecules, in a such way that, in living organisms, they become essential machinery in many aspects of DNA repair. Consequently, they have been engineered to target specific DNA sequences, by introducing cuts into their strands, enabling the removal of the existing sequence and the insertion of or replacement with others. Briefly, an enzyme cuts the DNA at a specific sequence, and when it is repaired, a change or 'edit' is made to the sequence.

Recently, a new promising technique has been introduced for gene editing, which has been hailed as the most important innovation in synthetic biology in recent years. While the common methods take a long time to edit gene sequences, thanks to the ease of use and accessibility of a new, cryptically named system called CRISPR/Cas, that time has been reduced by at least a few orders of magnitude. As shown in Fig. 9.2, it consists of two components, respectively:

– A guide RNA (gRNA),
– An associated nuclease, called Cas9, which is an enzyme that hydrolyses the phosphodiester linkages in RNA or in single- and double-stranded DNA, called, respectively, ribonuclease and deoxyribonucleases.

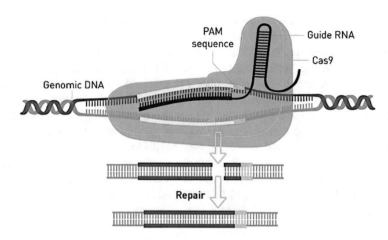

Fig. 9.2 The CRISPR/Cas system consists of two components, respectively, a guide gRNA and an associated nuclease, Cas9, which hydrolyses the phosphodiester linkages in the target DNA. For details see text

The gRNA is a short synthetic RNA composed of a sequence necessary for Cas9-binding and a user-defined sequence of about 20 nucleotides that specifies the genomic target to be modified. This genomic target of Cas9 can be modified by simply changing the targeting sequence present in the gRNA. The CRISPR/Cas9 mechanisms recognize DNA targets that are complementary to its short sequence, known as a protospacer. Then, in order for Cas9 to function, it requires a specific "protospacer adjacent motif (PAM)" that varies depending on the bacterial species of the Cas9 gene.

CRISPR was originally employed to "knock-out" target genes in various cell types and organisms, but modifications to the Cas9 enzyme have extended its application to the selective activation or repression of target genes, the purification of specific regions of DNA, and others. Actually, it is triggering a revolution, because many laboratories around the world are using such technology for innovative biological applications. Of course, its potential applications are very wide, as shown in the following scheme, inspired by an interesting review by two researchers who made an important contribution to its discovery, Jennifer A. Doudna and Emmanuelle Charpentier.

Future of CRISPR-cas 9-mediated genome engineering

- Human gene therapy
- Screens for drug targets
- Ecological vector control
- Viral gene disruption
- Agriculture crops
- Synthetic biology pathway engineering
- Programmable RNA targeting

But why call it "CRISPR", an intriguing name that recalls the crunchy chips that accompany our cocktails? Short for "clustered regular interspaced short palindromic repeats", it evolved in simple, single-celled microbes as a weapon for fighting off viruses. When viruses attack a bacterial cell, for instance, by injecting their own DNA, the cell responds by deploying CRISPR, that is, the strand of ribonucleic acid, hooked up to an enzyme associated with a protein, which is, by now, the familiar Cas9. But who has opened up this new frontier of research into synthetic biology that is presenting new opportunities in the field of genetic manipulation? The story, complex and fascinating, is described in great detail in the afore-mentioned thorough review published in 2014 in Science with the significant title *"The new frontier of genome engineering with CRISPR-Cas9"*. A second review with the epic title *"The Heroes of CRISPS"* was published in Cell in 2016 by Eric S. Lander; it is full of anecdotes about the scientists involved and the relevance of the work in progress.

9.4 Technical Issues

The capacity of biological systems to operate despite internal or external perturbations is called *robustness*. This outcome is mainly due to the intervention of cellular networks, which behave according to the feedback mechanism introduced in Chap. 5. Within this framework, the presence of bi-stable switches are fundamental for the regulation of complex systems, ranging from electronics to living cells. In order to gain insight into it, let us recall that the information-processing in electronic systems consists of a complex network of bi-stable electrical switches that implement logic functions operating according to Boolean algebra. Each switch has two stable steady-states, as illustrated in Fig. 6.2, so that at lower voltage, a binary '0' is present, while at high voltage, a binary '1' is present. To flip the switch from one to the other, the circuit is perturbed with a biasing voltage in one direction, and to flip it back, with a biasing voltage in the other direction. But what about a cell? Inside a cell, we can identify input signals and output responses, but in between, there is a jungle of bio-chemical reactions involving genes, mRNAs and proteins. Actually, as introduced in Sect. 6.4, a cell is basically working in a digital manner, because the basic logic of genetics is present within it. In other words, the principle of the logic gates introduced there, and familiar to anyone working in electronics, are also present in nature, as illustrated in 6.5. But where are the bi-stable switches? If a gene is activated, the encoded protein will also be activated, and it will then perform a function, as illustrated in Fig. 9.3, by considering a single-gene circuit that displays the fundamental building block present in circuits of higher complexity. mRNA molecules are synthesized from the template DNA strand at a rate of α_R, and proteins are translated at a rate of α_P of each mRNA molecule. γ_R and γ_P are the decay rates for mRNA and protein, respectively. Such networks can be compared to electrical circuits, because the state of a gene is characterized by ON or OFF states, depending on whether it is expressed or repressed.

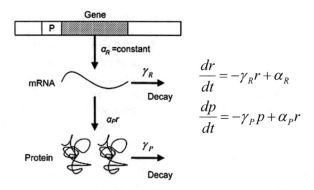

Fig. 9.3 A single-gene circuit that displays the fundamental building block present in circuits of higher complexity. mRNA molecules are synthesized from the template DNA strand at a rate of α_R, and proteins are translated at a rate of α_P of each mRNA molecule. $\gamma_R r$ and $\gamma_P p$ are the decay rates for mRNA and protein, respectively. *Simpson M, Cox CD, Peterson GD, Sayler GS (2004) Engineering in the biological substrate: information processing in genetic circuits. Proc IEEE 92 (5):848–863*

Researches have thus been conducted on the building of artificial networks able to mimic nature, with the perspective of attributing t novel functions to them. The occasion to do so was offered by the emergence of synthetic biology, in which the artificial networks appeared capable of being applied to the metabolic control. In practice, the construction of electrical circuits benefits from a large collection of well-characterized parts and modules, including resistors, capacitors, and inductors, connected together to generate circuits with useful functions. The progress in biotechnology is generating a similar list, specifically oriented towards facilitating gene-circuit design and implementation. This registry contains basic parts, including ribosome binding sites and transcriptional terminators. By taking advantage of it, many novel synthetic gene circuits can be created through a suitable design and its implementation. Of interest is the afore-mentioned synthetic gene circuit called the toggle switch, introduced by Timothy Gardner et al. in 2000. The name is due to its ability to program bi-stable behavior, corresponding to the afore-mentioned ON-OFF states. This is due to the interaction between the actions carried out by substances that activate the possible transformations and others that repress them. The corresponding mathematical modeling includes the transient material balances of the involved components, according to the material balance equation. The steady-state solution is given by a couple of algebraic equations (Box 9.1), obtained by putting the derivative of the concentrations with respect to time as equal to zero. For appropriate values of the circuit parameters reflecting the values of the rate constants, the presence of bi-stability comes out. In other words, for particular parameter sets, either the stable high state or the stable low state can be reached, depending on the prior history of its dynamics. The model predicts the conditions under which bi-stability is favored, by offering the possibility of guiding the choice of the repressors.

Another known circuit, called the repressilator, was introduced in the same year by Michael Elowitz and Stanislas Leibler. It consists of a network of three sections of DNA sequentially repressing each other, akin to the rock-paper-scissors game. In it, a negative feedback loop with a long cascade is present, where the first repressor protein (TetR) inhibits the second repressor gene, whose protein product (lcI), in turn, inhibits the third repressor gene (lacI), and finally, the third repressor protein (LacI) inhibits the first repressor gene. This network was designed to exhibit stable oscillations, and hence it acts like an electrical oscillator system with fixed time periods. The wave continues as each gene inverts the output of the next, so that the ON-OF behaviour is present in an oscillating wave. The network was implemented in *Escherichia coli* using standard molecular biology methods, and observations were performed that verify that the engineered colonies do indeed exhibit the desired oscillatory behavior.

9.5   Works in Progress

A wide spectrum of research activities in the field of synthetic biology is emerging from that which preceded it, with an extraordinary potentiality in its applications, including, as a relevant perspective, the creation of a toolkit of functional units that, when introduced into living cells, could give rise to the following new technological functions:

– The achievement of a form of genome engineering, focused on the construction of synthetic genomes for whole or minimal organisms. Several key technologies are, of course, critical to the development of such a program; among them, the advancement of the omics techniques introduced in the Chap. 4 occupy the primary position, by adding the standardization of the biological entities and the acquisition of the skills imposed to reach the creativity required in their application to increasingly complex synthetic systems.
– The improvement of the modeling of engineered biological systems able to predict systems' behavior prior their fabrication, particularly for that which concerns the ways biological molecules bind onto substrates and catalyze reactions, the way in which DNA encodes the information concerning the cells, and finally, the way in which multi-component integrated systems behave.

Given the above, the picture of the potential applications of synthetic biology is impressive. Only some of them will be mentioned.

Industrial enzymes. Researchers and companies aim to synthesize enzymes with high activity, so as to obtain products with optimal yields and effectiveness. This occurs within the frame of a fallout of synthetic biology, known as metabolic engineering, which can be considered a biotechnological technique utilized in industry to discover pharmaceuticals and fermentative chemicals.

Unnatural amino acids and nucleotides. One topic of investigation is the expansion of the normal repertoire of the twenty amino acids, which are coded into all organisms. Some of them are engineered to code for alternative non-standard amino acids'. Moreover, technologies have been developed for incorporating unnatural nucleotides and amino acids into nucleic acids and proteins, both in vitro and in vivo.

Material production. By integrating synthetic biology approaches with material science, it would be possible to produce materials with properties that can be genetically encoded. Nanofibers have been constructed for specific functions, including: adhesion to substrates; nanoparticle templating; and protein immobilization.

Biosensors. This refers to engineered organisms, usually bacteria, that are capable of reporting some ambient phenomenon, such as the presence of heavy metals or toxins.

Information storage. Scientists can encode vast amounts of digital information onto a single strand of synthetic DNA. For instance, Georg Church was able to encode one of his books about synthetic biology onto DNA.

Biological Computers. This topic concerns the engineered biological systems that can perform computer-like operations. Researchers are focused on the building and characterization of a variety of corresponding logic gates, as well as both analog and digital computation activities that are present in living cells.

Cell transformations. At present, entire organisms have not yet been created from scratch, following the dream of Leduc. Nevertheless, living cells are being transformed through the insertion of new DNA, and even entire synthetic genomes, which, once integrated into a living cell, is expected to manifest the desired new capabilities of growing and thriving.

Synthetic life. Of course, one of the most important, if not *the* most important, topic is the creation of artificial life in vitro from biomolecules. This bottom-up approach, reminiscent of the Leduc program, relies on the conviction that fundamental concepts about life, particularly self-replication, can be chemically implemented. Clearly, there is continuity with the research program of prebiotic chemistry, as established by Stanley L. Miller and Harold C. Urey, leading to the ultimate knowledge of the nature of life.

9.6   A Renewed Synthetic Chemistry

Even though the impact could be enormous, the potentiality of living cells, to date, has been only partially utilized in chemical production activities. In fact, the sophisticated circuitry present in the bacterial enzymatic system opens up new stimulating options for the achievement of synthetic processes that can produce chemicals. As a matter of fact, through the improvement of the metabolic cellular networks, new bioprocesses could be implemented, able to achieve the synthesis of complex chemicals. The employment of biological organisms to transform precursor molecules into targeted products dates back to the earliest days of human history, in the form of fermentation to obtain beer, cheese, and bread. Nevertheless, the modern era of industrial biotechnology began more recently, through the development of the large-scale fermentation of penicillin. Even though the molecule was isolated by Sir Alexander Fleming in 1928, it was not produced on a large scale until the Second World War. It is interesting to observe that such achievements preceded the elucidation of the structure of DNA by nearly a decade. In the following years, the fermentation processes were developed for large-scale commercial production of several products, including citric acid, glutamic acid, and others. Herbert Boyer, Stanley Cohen and their colleagues, in 1973, published a paper entitled "*Construction of Biologically Functional Bacterial Plasmids in Vitro*". In this publication, the first genetic engineering experiment was described, so that the new biotechnology took off. The occurrence of biochemical synthesis as the result of a series of enzyme-catalyzed reactions taking place in complex pathways creates the challenge for a more systematic approach within the framework of system science. The emerging challenges that needed to be addressed were clarified in a paper published in 1992 by James E. Bailey, entitled "*Toward a Science of Metabolic Engineering*". The new science, baptized *Metabolic engineering*, while extensively applying the emerging omics analytical procedures for network analyses, sought to take advantage of the tools of recombinant DNA technology, modifying them in order to obtain more effective strains. These principles were applied so as to generate efficient and productive fermentation processes.

In reality it can not be forgotten the fact that the development and the applications of such new technologies must reserve a careful attention on their fallout on society. This topic has been critically discussed in a recent EMBO report (June 2018) written by Benjamin Trump and others. It starts from the awarness that normally the researches developed by men are growing faster than their social repercussions and their applications. Concerns of risks due to synthetic biology up to now have been focused on biosafety, mainly jeopardized by accidental releases, and by the misuse of the previously mentioned technologies. Further concerns are connected with the risks due to the proliferation of mutated engineered organisms. In conclusion, it comes out that in synthetic biology the sinergy between social and physical sciences is a prerequisite to consolidate the underlying interdisciplinary approaches. Only it can guarantee a transparent and responsible aptitude on the best practices to follow in order to control the risk concerns.

9.7   Metabolic Engineering

The networks of chemical reactions catalyzed by enzymes controlled by genes can be compared to chemical engines that support cellular functions by converting raw materials into energy and into the molecules required to build the biological structures. In fact, cells are capable of synthesizing a great array of chemicals under gentle conditions that are aqueous solutions at mild temperatures. Conversely, the current methods for organic synthesis often rely upon the employment of exotic solvents and expensive catalysts, sometimes operating at high temperature and pressure. Such a situation justifies the interest in a new catalytic approach that takes advantage of the progress in synthetic biology. In it, the microbial organisms are utilized as cell factories for the bioconversion of renewable resources into bulk or high value chemicals. In an engineering approach, the concept of the chassis is introduced, referring to a framework (internal in bacteria) that accommodates a synthetic system, including the energy sources, as well as the transcription and the translation machinery typical of metabolic environments. The introduction of novel production pathways is the core of a new approach that combines biology with engineering. Because the metabolic pathways are often branched, it is possible to switch from one route to another by modifying the activity of one, or more, enzymes that act as a rate-determining step of the pathway:

$$\textbf{substrate} \xrightarrow{\text{enzyme 1}} \textbf{product A} \xrightarrow{\text{enzyme 2}} \textbf{product B} \xrightarrow{\text{enzyme 3}} \textbf{final product}$$
$$\downarrow$$
$$\textbf{alternative product}$$

The alternative product can become the starting substrate for another pathway. The focus is to individuate and then characterize the key enzymatic reactions able to control the pathway flux and then to modulate their activities to favor the formation of the desired products. The controls, which are generally fast (i.e., seconds to minutes) and inexpensive energy-wise, are made by means of regulatory procedures that modulate the activity of the preexisting enzyme molecule. The simplest approach is the modification of the concentration of the substrate S, because the rate of an enzyme-catalyzed reaction follows the afore-mentioned catalytic kinetics of Michaelis–Menten and then depends hyperbolically on the concentration of S. Not all enzymes follow the simple Michaelis–Menten rate law, because, as mentioned in Chap. 4, they often contain more than one binding site, and thus, as shown in Fig. 4.7a, the non-covalent binding with a specific metabolite (dark square) can affect their conformation and positively enhance the binding with the substrate. Enzymes of this nature, called allosteric, can assume other shapes, and as a result of such a cooperative binding, a sigmoidal relationship between enzyme activity and the concentration of the substrate emerges that leads to the typical hyperbolic input/output relationships illustrated in Fig. 4.7b. Computation is thus becoming a crucial ingredient when dealing with the description of the metabolic functions of a cell, and therefore the simulation in silico of the cells is a challenge in catalytic reaction engineering, because the corresponding mathematical model can be regarded as a virtual laboratory that gives insight into the cellular functions. The electronic systems are manufactured within substrates where information, represented by a collection of charge carriers, is processed through the control of their transport between circuit nodes. In a similar way, the information in the cells is represented by a collection of molecules that is processed through the control of their synthesis, decay and transport. In electronic systems, the transport of charge carriers depends on the electrical potential difference between connected nodes, while the chemical potential plays an analogous role in determining the rate of information transfer in the biological substrates. The gene is the fundamental unit of memory of the cells, due to the sequence of nucleotides in their DNA. It is read, or expressed, when it is opened into a single strand. It is the fallout of the biotechnological revolution that is greatly expanding the perspectives of metabolic engineering by taking advantage of the rapid decline in the cost of commercial gene synthesis, a phenomenon similar to Moore's law for electronics. This is a good auspice, because when a technology becomes similar to an informatic technology, it begins to evolve exponentially.

An interesting example of work in progress in metabolic engineering concerns the exploitation of the properties of the familiar cyanobacteria. The phylum of bacteria, often called blue-green algae, which obtain their energy through photosynthesis by using a specific type of chlorophyll together with other pigments, are able to acquire energy from solar light. The energy is used to enzymatically convert carbon dioxide from the air into the nutrients needed for the growth, by generating oxygen as a byproduct. Therefore, cyanobacteria are a promising biological chassis for the synthesis of renewable chemicals and fuels, thanks to their genetic plasticity and with their wealth of information comprising more than 170 genome sequences determined up to now. The researches are addressed at introducing enzymes from different organisms into cyanobacteria in order to construct biosynthetic pathways for targeted compounds. A number of valuable compounds have already been produced (e.g., ethanol, isobutyraldehyde, isobutanol, 1-butanol, isoprene, ethylene, hexoses, cellulose, mannitol, lactic acid, fatty acids), as shown in Fig. 9.4.

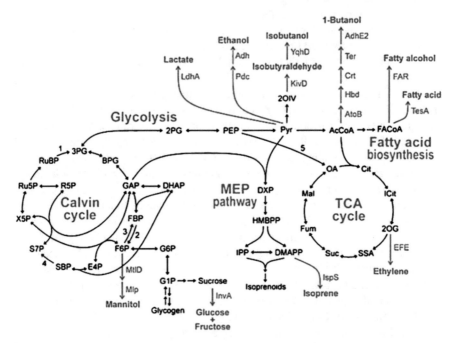

Fig. 9.4 Valuable compounds that have already been produced in cyanobacteria (e.g., ethanol, isobutyraldehyde, isobutanol, 1-butanol, isoprene, ethylene, hexoses, cellulose, mannitol, lactic acid, fatty acids). *Knoop H, Steuer R (2015) A computational analysis of stoichiometric constraints and trade-offs in cyanobacterial biofuel production. Front Bioeng Biotechnol 3:1–15*

9.8 Towards a Post-petroleum Era?

Starting 300 years ago, the traditional wind- and water-powered technologies were replaced by coal and hydrocarbons, which, at present, cover more than 80% of overall energy consumption. In the meantime, hydrocarbons became a primary source of polymeric materials, detergents and fertilizers. At the beginning of the last century, the alliance between oil and the car took off, a dizzying development that have rise to one of the largest existing production activities on the planet. The appropriate mixtures of hydrocarbons used as fuels are obtained by oil through reaction and separation processes, integrated into the complex structures of the refineries. A petroleum refinery includes catalytic processes that concern the transformation of the hydrocarbon molecules into more complex ones with a higher octane rating value. These processes are performed in the afore-mentioned fluidized bed reactors, which can be considered a hybrid between a packed bed reactor and a stirred reactor. The exploitation of oil underwent a qualitative and quantitative leap in the second half of the last century, as testified to by the gigantic dimensions of the present refineries, which have the capacity to process daily more than half a million barrels of oil daily. This rapid development, helped by the identification of new catalytic processes, mostly based on synthetic zeolites, is able to transform the hydrocarbon molecules into others more appropriate for combustion management. Thanks to catalysis, fuels with high performance are obtained, devoid of heteroatoms (sulfur, nitrogen), which act as pollutants. However, what about the carbon dioxide emission? The accepted projections on greenhouse gas emissions in the twenty-first century indicate that the stabilization of the level of carbon dioxide concentration in the atmosphere can be pursued only through the introduction of carbon-free technologies, particularly because the gas emissions resulting from transportation will reach a level beyond which it will be impossible to put a reasonable target on carbon dioxide concentration. Unfortunately, the global shift away from fossil fuels towards an extensive electrification, able to cover the scale of present energy consumption, implies an enormous infrastructural requirement. In this context, it is enlightening to mention the following sentence from a biologist: "*Many other fields of science and engineering have developed system science and complicated mathematical simulation. Chemicals originate from highly integrated chemical processes with structures that rival those of living cells, designed in a computer. What about biology? Vast amounts of basic genetic and biochemical information are rapidly becoming available. Sequencing technology is providing us with complete information about the genetic makeup of simple cells, and more complex organisms will soon follow*" (B. P. Palsson, Nature America 2000). Such an awareness supports the approach of relying on the ability of microorganisms to use renewable resources for biofuel synthesis. In it, microorganisms are explored for the production of a next-generation of biofuels, by taking advantage of the possibility of modifying their metabolic networks (Liao et al.). One of the advantages of using microorganisms for the production of a next-generation of biofuels is the metabolic diversity of bacteria, fungi, and microalgae, which enables the use of different substrates as the starting point for biofuel generation. New approaches are in progress, including ones that employ the ability of bacteria and fungi to utilize lignocellulose from wastes and energy crops, the ability of bacteria and microalgae to fix CO_2 and the capacity of microorganisms to use methane from landfill or natural gas wells that would otherwise be flared.

Box 9.1 Toggle Switch

The circuit of the toggle switch consists of a complex system in which two chemical compounds, TetR and LacI, mutually repressing each other, are present. The circuit can be flipped between "high" and "low" states using chemical inducers, specifically IPTG (*isopropyl-b-D-thiogalactopyranoside*) and aTc (*anhydrotetracycline*). For example, when aTc is added to the system to inhibit TetR, LacI will be highly expressed and the exit will be OFF. Conversely, when IPTG is added to inhibit LacI, TetR will be highly expressed and the exit will be ON. Such conditions for bistability are simulated by using the following dimensionless mathematical model derived by the dynamic material balances of the two repressors:

$$\frac{du}{dt} = \frac{\alpha_1}{1 + v^\beta} - u$$
$$\frac{dv}{dt} = \frac{\alpha_2}{1 + v^\gamma} - v, \tag{9.1}$$

where u is the concentration of repressor 1, v is the concentration of repressor 2, α_1 is the effective rate of synthesis of repressor 1, α_2 is the effective rate of synthesis of repressor 2, and β and γ are parameters that account for the action of promoter 2 and 1, respectively. The two most fundamental aspects of the network are preserved, respectively, in the cooperative repression of promoters (the first term in each equation) and the degradation/dilution of the repressors (the second term in each equation).

References

Benjamin D Trump, Jeffrey C Cegan, Emily Wells, Jeffrey Keisler and Igor Linkov, A critical juncture for synthetic biology EMBO reports 19: e46153 I 2018

Carlson Robert H. *Biology is Techonology*, Harward Uniersity Press, 2011.

Church George, Ed Regis *Regenesis*, Basik books, 2012.

Doudna, Jennifer A, Emmanuel Charpentier. *The new frontier of genome engineering with CRISPR-Cas9*, Science, 346,n.6213,1077, 2014.

Reeves Gregory T, Curtis E. Hrischuk. *Survey of Engineering Models for Systems Biology Computational,* Biology Journal Volume 2016, https://doi.org/10.1155/2016/41

Carrà Sergio. *Peculiarity and perspectives of catalytic reaction engineering*, Rend. Fis. Acc. Lincei DOI https://doi.org/10.1007/s12210-017-0598-y.

Industrialization of Biology: A Roadmap to Accelerate the Advanced Manufacturing of Chemicals, THE NATIONAL ACADEMY PRESS, WASHINGTON, D.C., 2015, www.nap.edu.

Chubukov, Victor, Aindrila Mukhopadhyay, Christopher J Petzold, Jay D Keasling and Héctor García Martín. *Synthetic and systems biology for microbial production of commodity chemicals.* npj Systems Biology and Applications (2016) 16009, 2016. The Systems Biology Institute/ Macmillan Publishers Limited.

Chapter 10
Why Life?

10.1 Resurrecting Prometheus

In 1816, the English writer Mary Shelley wrote the Gothic novel "*Frankenstein, or A Modern Prometheus*", inspired by the Titan from Greek mythology best known as the bringer of fire to humanity. In the book, the story of Victor Frankenstein, a young scientist who produces a grotesque but sapient creature in an unorthodox scientific experiment, is told, taking the occasion to express some reflections on the possibility and opportunity of creating an artificial life. At that time, a harsh cultural controversy was afoot, focused on the conflict between vitalists versus mechanicists. The former were arguing that living things are distinguished from matter by some animating principle, baptized the "*spark of life*", while the latter compared the animals to clockwork automata in compliance with a philosophical view supported by the mechanistic physiological theories of Cartesio.

At present, at least in the scientific community, the doctrine of vitalism has been dropped, but nevertheless, there is still resistance to acceptance of the idea that life can be understood by disassembling an organism into its component parts. This possibility, however, has been reinforced by the occurrence of the "eighth day of creation", upon which it was discovered that DNA is the blueprint of life, being comprehensive of the full set of instructions required for building a cell. In the meantime, the acquisition of omic-scale data, as mentioned in Sect. 4.7, is yielding a vast amount of basic genetic and biochemical information, providing us with extensive knowledge about the genetic makeup of existing cells. The progress in the simulation of a living cell with a computer program can be considered the first step in such a direction, but nevertheless, despite some important advances, it appears that the mere knowledge of the existence and role of DNA is not yet the master key to life. In other words, are we really on the eve of the fulfillment of a synthetic approach to biology, capable of producing cells with well-defined characteristics, and therefore useful to be employed with confidence in the development of man-made technologies?

© Springer Nature Switzerland AG 2018
S. Carrà, *Stepping Stones to Synthetic Biology*, The Frontiers Collection,
https://doi.org/10.1007/978-3-319-95459-2_10

In 2010, a team of researchers from the Craig Venter Institute announced the obtainment of the first synthetic cell starting from a genome made from chemicals transplanted into the recipient membrane of an existing cell. This achievement represented a new and significant level of control over living matter at the molecular level. The researchers involved, by emphasizing the relevance of their achievement, recognized that it is only the first step towards the engineering of life, despite the significant progress obtained in the fundamental understanding of cell biology.

As a matter of fact, some important conceptual issues have yet to be faced, concerning the way in which cell biology is constrained by the laws of physical-chemistry, by deciphering the way they channel the cellular characteristics into a limited range of alternatives. It is thus important to bridge the gap between the fields of cell biology and evolutionary biology within the frame of the principles of biochemistry and physical chemistry. Researches suggest that the key first steps in the origin of life are inevitable consequences of the behavior of organic molecules in water, so that the characteristics of the complex molecular structures emerging spontaneously from their interactions become of paramount importance. Thus, the time is ripe to approach the following questions:

- When did life start on the earth?
- What do we know about the beginning of life on the earth?

10.2   The Beginning of Life

Within the framework of the cosmic evolution presented in the Chap. 3, Earth was born 4.54 billion years ago, approximately, because everything happened in a molten phase, starting from a solar nebula and involving frequent collisions with other bodies, which led to volcanism associated with the outgassing processes that probably created the primordial atmosphere and the oceans. This occurred with a small amount of oxygen, subsequently creating an environment unsuitable to present life. The geological clock indicates the existence of a large span of time before the appearance and evolution of life on the planet. Even though the earliest reported fossil discoveries date from 3.5 billion years ago, it wasn't until approximately 600 million years ago that complex multi-cellular life began to enter the fossil record, because most living organisms simply decayed without leaving traces. Actually, single cellular bacteria existed from very early in the history of life on Earth, because there are reasonable arguments that they were present a[?] billion years ago, even though it is recognized that their significant spread started roughly 1.8 billion years ago, when oxygen appeared in the atmosphere as a result of the action of cyanobacteria. Bacteria have thus had plenty of time to adapt to their environments by giving rise to numerous descendant forms. The fossils found in ancient rocks show a progression from simple to complex prokariotic single-celled organisms, resembling present-day bacteria, of 1 to 2 micrometers in diameter, but eukaryotic bacteria did not appear until about 1.5 billion years ago. Therefore, for at least 2 billion years, nearly half of the age of the earth, prokariotic bacteria were the only organisms present.

In the oxygen-free places, such as the depths of the Black Sea or the boiling waters of hot springs and deep-sea vents, some bacteria are still living at very high temperatures without oxygen. Some are producing methane from CO_2 and H_2. Called archaebacteria as a whole, they resemble all other bacteria in having DNA, a lipid cell membrane, an exterior cell wall, and a metabolism based on ATP. Of course, biologists have attempted to categorize similar organisms, but cutting edge work was published in 1977 by Carl Woese and George Fox, both at the University of Illinois. In it, all monocellular organisms present in the world were arranged into three categories on the basis of the structure of their ribosome, the familiar complex nanomachines composed of both proteins and RNA. The resulting classification includes the following categories:

Archaea: prokaryotic organisms that lack a cell wall, including the methanogens and extreme halophiles and thermophiles.

Bacteria: prokaryotic organisms with a cell wall, including cyanobacteria, soil bacteria, nitrogen-fixing bacteria, and pathogenic (disease-causing) bacteria.

Eucaria: Eukaryotic, primarily unicellular, although algae are multicellular, photosynthetic or heterotrophic organisms, such as amoebas and paramecia.

Thanks to his detection of archaea, Carl Woese is considered the microbiologist who uncovered the "third domain" of life, by redrawing the identification and classification criteria of organisms and proving that all life on Earth is related. As illustrated in Fig. 10.1, according to a new taxonomic tree, the two types of prokaryote, called Bacteria and Archaea, diverged from a common precursor very early, even though a prominent difference was that only bacteria widely adapted to aerobic conditions. Within this frame, quite interesting and important is the fact that Woese supported the Darwinian idea that all life on the planet is connected to a remote unitary ancestor called LUCA, an acronym for Last Universal Common Ancestor. Actually, starting witj Darwin, biologists have fostered the belief that all living things share a common origin, but the convinction was reinforced by the discovery of the genetic code, the apparatus for the transcription of genetic information. In this awareness, LUCA is at the first node of the line of descent leading to archea and eukaria from that of bacteria. Actually, Woese envisaged the presence of a miscellaneous primeval community of protocells able to exchange primary unfixed genes, but evolving as a unit by assuming successful configurations, until the emergence of modules able to process the genetic information.

But what about the emergence of oxygen if the fossil records indicate that terrestrial plants appeared on the earth about 450 million years ago? Of course, the advent of photosynthesis was one of the central events in the early development of life on Earth, however, its origin and evolution have long remained unresolved. Even though the picture is not altogether clear, it is reasonable to assume that some prokaryotic microbes have evolved complex molecular machines able to split water by using solar energy a billion years before the rise of terrestrial plants. At present, the only extant group of photosynthetic microbes capable of producing oxygen are the cyanobacteria, which are small aquatic bacteria, usually unicellular,

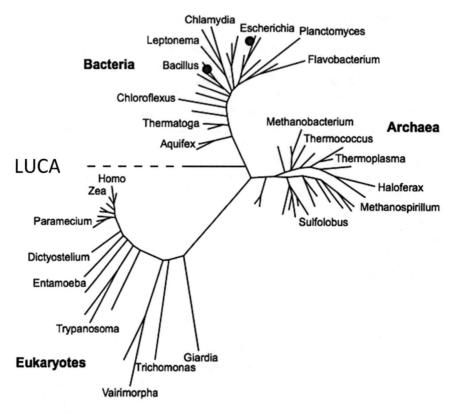

Fig. 10.1 The phylogenetic tree of life, showing the representative species from three kingdoms. Accordingly, two types of prokaryote, called Bacteria and Archaea, diverged from a common precursor very early on. *Pace NR (1997) A molecular view of microbial diversity and the biosphere. Science 276: 734–740*

that also grow in colonies. Notwithstanding the fact that their evolution is still enigmatic, some studies have demonstrated that the present photosynthetic eukaryotic cells acquired their properties from the endosymbiosis of an archean host cell with a cyanobacterium. As a matter of fact, the present oxygenic photosynthetic machine, as described in Sects. 6.6 and 6.7, is the most complex energy-transducing apparatus present in nature, which evolved only once in the history of the earth. Cyanobacteria became the first microbes to produce oxygen on the earth through photosynthesis, perhaps as long ago as 3.5 billion years, but, mysteriously, it required a long lag time, about 1 billion years, before the atmosphere first gained significant amounts of oxygen. As to the moment at which cyanobacteria started to produce oxygen with an impact on the atmosphere sufficient to promote the emergence of animals, it seems probable that such an event happened about 2.3 billion years ago.

10.3   Ideas and Theories on the Origin of Life

10.3.1   Protein First

In the investigations into prebiotic evolution, three phases can be distinguished, defined as geophysical, chemical and biological, respectively. The first concerns the study of the primitive conditions that existed on Earth, with particular reference to the composition of the atmosphere, while the second concerns the chemical processes through which the synthesis of the organic molecules constituting the bricks of biopolymers, namely amino acids, sugars, azotated bases and others, occurred. These two phases are well-understood. Certain aspects of the chemical phase have been reproduced in laboratories, through the creation of an environment similar to the one that, in agreement with geophysical theories, existed at the dawn of life. It is believed that at that time, the earth's atmosphere was oxygen-free, according to a hypothesis first advanced by J.B. Haldane in 1929 and subsequently confirmed by geologists and paleochemists. The uncertainties concern the biological phase, in which life would have appeared in a "primordial soup" present on the planet. In fact, alternative interpretations and some perplexities are still hanging on the occurrence of such a singular event.

Let us start our survey on the theories on the origin of life from the pioneering work of the biochemist Alexandr Oparin, a member of the URSS Academy of Science, who, in 1936, published a book on the subject. In it, a primary role was attributed to droplets of organic materials as precursors of cells. In other words, life started to develop from microscopic, spontaneously-formed spherical aggregates of lipid molecules, called coacervates, held together by electrostatic forces. In such a context, the concept of self-replication as a collective property of molecular assemblies was clearly stated as follows:

> ...the stationary drop of a coacervate, or any other open system, may be preserved as a whole for a certain time while changing continually in regard to both its composition and the network of processes taking place within it, always assuming that these changes do not disturb its dynamic stability. ... This constant repetition of connected reactions, co-ordinated in a single network, also led to the emergence of a property characteristic of living things, that of self-reproduction. This may be taken as the origin of life.

Quite important is the fact that such a scenario, subsequently called "metabolism first", was established without accounting for the possible role of nucleic acids, additionally because their relevance and structure were unknown when Oparin introduced the theory. Nevertheless, the acquisition of the awareness of the role played by chemical processes in the origin of life fostered some of the subsequent developments aimed at justifying the occurrence of the self-replication processes in molecular mixtures. The prerequisite was the presence of collections of molecules in a state able to promote mutual catalytic reactions between the different components. Thanks to the fluxes of energy and the molecules coming from the environment, such assemblies can expand by increasing their catalytic mixture until the emergence of a selection process capable of triggering an extensive evolutionary process.

Physical support of the previous theory came only several years after its formulation, from Freeman Dyson, who, in 1985, developed a toy model able to justify the transitions from a poorly organized system to a network of well-connected catalytic interactions. In it, it was assumed that the molecules enclosed in a bounded micro-environment are present in two states, catalytically active and catalytically inactive, respectively, whereby the active molecules could convert the other molecules into a similarly privileged state. From the chemical point of view, attention was focused on the processes through which the macromolecular proteinic molecules are formed from the molecules of amino acids confined within the afore-mentioned droplets acting as microreactors. Different states can be established in correspondence with different levels of organization associated with the degree of metabolic activity exerted by the macromolecules. In fact, the catalytic activity depends on their configuration, because a different active role can be attributed to the different monomers, in correspondence to their position in the polymer molecules. In the approach by Dyson, the molecules have no specific identity or structure, but only a functional property expressed through their catalytic power. In other words, from a physical point of view, a mean field approach was adopted by assuming that the description of the behavior of a mixture of catalytic molecules can be approximately faced by attributing to any specific molecules the average value of the catalytic activity of the entire mixture. Attention must then be focused on the behaviour of the probability $\varphi(x)$ that an inactive monomer will be activated as a function of the fraction x of the active monomers present in each droplet. If it is assumed that changes in the molecular population are due to the replacement of a monomer with another at a particular point of a polymer, the S-shape curve shown in Fig. 10.2 can be derived. A steady-state occurs when the actual values of the concentrations no longer change, as explained in Sect. 2.3, that is, whenever $\varphi(x) = x$, because under such conditions, the daughter population inherits the same average activity x as the parent population. In a Cartesian plot, the curve $y = \varphi(x)$ cuts the diagonal at the three points, which are in correspondence with three possible steady-states of the system, indicated with α, β, γ, respectively. α and γ are both stable states. In fact, for instance, in correspondence with the state γ, if x is increased, the generated molec-ular population has a lower activity than the one corresponding to the initial stationary state. Therefore, the system backs off spontaneously to the point γ. If, instead, the value of x decreases, the activity is higher, and therefore the system also returns in this case to the stationary value. With similar analysis, it can be shown that the point α is, in turn, stable, while the point β is unstable. In conclusion, the system has three stationary states: two extremes, both stable, and one unstable intermediate. Their level of organization depends on the $\varphi(x)$ value, because if it is low, it does not turn out to be compatible with the necessary life-generating process that requires a high level of catalytic activity. Therefore, it depends on the possibility that the system can jump spontaneously from α to γ. This occurrence can happen only stochastically, that is, when the number of amino acids present in a drop is not very high. Dyson deepened the analysis by determining the values that must be attributed to model parameters, by obtaining a set of values compatible with the

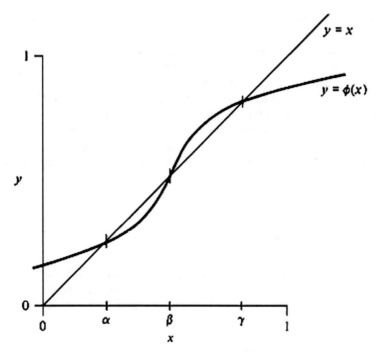

Fig. 10.2 The S-shaped curve $y = \phi(x)$. The curve crosses the line $y = x$ at three points that represent possible stationary states of the populations. The upper and lower states are stable, and the middle state is unstable

characteristics of a natural amino acid mixture, therefore supporting the Oparin model.

In the spirit of enlarging the typology of the molecular mixture, by also including some nucleic acids, in 1993, Stuart Kauffman, working at the Institute of Complex Systems of Santa Fè, adopted a different strategy. He considered a mixture of monomers, dimers, trimers and so on, including peptides, RNA sequences and other polymeric forms. The number of monomer units is about 20 for proteins and 4 for nucleic acids. Then, he placed attention on all of the ruptures and bond formations through which the mixture is transformed. In a statistical approach, a constant value is attributed to the probability that a molecular species will catalyze any reaction, leading to the formation of another species. Kauffman, taking advantage of the employment of the cellular automata approach, was able to show that any sufficiently complex set of molecules may attain an emergent collective property, called catalytic closure, by which any reaction that must find a catalyst does find it. Further explorations confirmed that the considered mutually catalytic set can represent a candidate for a primordial unit on which natural selection could act.

In conclusion, both of the previous results give us a glimpse at the possibility that complex catalytic polymer systems can experience sudden phase transitions to configurations in which the presence of a network of chemical interactions could establish a high level of organization. In such a framework, the origin of life would result by taking advantage of the self-organizing properties of complex chemical systems. However, the hypothesis, while being suggestive, neglects the role of the information that, as shown in Chap. 7, constitutes a necessary prerequisite to transmit, in processes occurring through a succession of stages, the memory of the characteristics that the system reaches along its evolution. At the molecular level, this role is carried out by the nucleic acids, so that their presence in any model concerning the origin of life cannot be neglected. But another aspect that needs to be dealt with concerns the mechanisms by which energy has contributed to the development of the afore-mentioned processes through the revamping the role of thermodynamics, whose implications cannot be ignored.

10.3.2   Nucleic Acids First

In 1971, a few years before the publication of Dyson's work, Manfred Eigen, Nobel prize laureate for his researches into the kinetics of very fast chemical reactions, published a theory on prebiotic chemistry, an alternative to that of Oparin. In it, the following succession of the involved elements was assumed:

GENES → ENZYMES → CELLS

The idea was borrowed from the discovery of the twin helix structures of DNA, which appeared, in some respects, more suitable than that of the proteins, as suggested by the famous last sentence of Crick and Watson's publication: "*It has not escaped our notice that the specific pairing we have postulated immediately suggests a possible copying mechanism for genetic material*". The Eigen approach provides a model to describe the synthesis of nucleic acids from some components present in the primordial soup, through a sequence of reactions capable of extracting biopolymers from the reacting mixture. To fulfill these conditions, the process must be characterized by a strong selectivity able to drive the emergence of particular biomolecules. In order to fulfil such a condition, Eigen introduced a molecular evolutionary scheme borrowed from Darwinian theory, articulated around the concepts of variation, selection and competition, as applied to biological macromolecules. The reagent system is compared to a continuous chemical reactor, in which an aqueous solution of monomers and macromolecules is present. The system is open to the exchanges of matter and energy with the surrounding environment, being permeable to the monomeric molecules rich in energy, which are transformed through a sequence of reactions into polymers. In order that a self-organizing process can take place, autocatalytic mutation and reproduction processes must be operative, leading to the formation of new types of biomolecule. The evolution of the system is described through a set of differential equations, each expressing the variation in time of the concentration of every polymeric species. For a polymer, consisting of N units of nucleotides, each indicated by i, whose concentration is expressed through their molar fractions x_i, each of the mentioned set of differential equations is written as follows:

$$\frac{dx_i}{dt} = (A_iQ_i - D_i - \Phi)x_i + \sum_{i \neq j}^{N} \varphi_{ij}x_i. \tag{10.1}$$

The first term in brackets on the right-hand side is called the selective value of the i-th class of the polymer. A_i is a parameter that characterizes the autocatalytic process through which the macromolecules with i sequences, regardless of their disposition, are generated per unit time, while Q_i is a number included between zero and one, which indicates the fraction of the produced molecules that are copies of the original one. The parameter D_i, instead, expresses the disappearance of the macromolecules of the class i by decomposition, while φ_{ij} is the mutation rate, which can lead to transitions to i molecules from j molecules. $\langle E \rangle = \sum_{i} (A_i - D_i)x_i$ is the average productivity of the different polymers, and finally, Φ is an outflow term that takes into account the removal of molecules from the system. Obviously, there is a similar equation for each sequence i, and on the whole, a system of non-linear ordinary differential equations is obtained. In each of them, the first term on the right-hand side may be positive or negative, depending on whether W_i is bigger or smaller than the average production $\langle E \rangle$. In the first case, x_i grows, while in the second case, it decreases towards the extinction of the sequence itself. This fact is a consequence of a tight competition that brings one of the sequences to prevalence over the others. The previous set of equations gives us a description of the population dynamics of a reactive system, characterized by the following properties:

- **Metabolism**, expressed by the first right-hand term, which gives the turnover from energy-rich into energy-deficient compounds.
- **Self-reproduction**, because the rate of formation of a molecular species i is proportional to its concentration.
- **Mutability**, as expressed by the factor Q_i.

Calculation conducted on simple systems of a few competing molecules revealed that by starting with the same concentrations of the initial components, as time passes, one species prevails over the others. In fact, two effects in competition are present in the model, as it appears from the right-hand side of Eq. (10.1), in which, owing to the presence of the term A_iQ_i, the sequence with its highest value will displace all of the others by suppressing the role of the mutations, despite the fact that such a contribution in an evolutionary process cannot be discharged.

With such awareness, Eigen focused his attention on the fact that the length of the macro-molecules obtained from his model was hindered by the copying fidelity, because it put a limit on the permissible size of the polymer. In an approximate analysis, it turned out to be worthwhile to employ the geometric mean q of the parameters Q_i. Its value, included between 0 and 1, expresses the probability that a single nucleic base is replicated correctly. It comes out that the number N of nucleotides present in a polymeric molecule is inversely proportional to $(1 - q)$, so that as q approaches 1, the polynucleotidic molecules become increasingly bigger. Unfortunately, the values of q occurring at the origin of life are low, because they correspond to non-enzymatic replications, so that values of N no higher than 100 are obtained. In order to increase the size of the macromolecules, the presence of enzymes is necessary. Laconically, the result can be summarized as follows: no large genome without enzymes, no enzyme without a large genome. This finding has been called the catch-22 of prebiotic evolution, borrowing the expression coined by Joseph Heller in his well-known novel of 1961, to indicate a paradoxical situation from which an individual cannot escape because of contradictory rules.

The validity of the Eigen model was first experimentally confirmed in 1967 by Sol Spiegelman, an American molecular biologist who developed the technique of nucleic acid hybridization, and later by the Eigen again, together with Pieter Shuster. As a reference system, RNA was chosen, which, under appropriate conditions, is able to replicate itself. The environment consisted of a solution containing the same nucleotide monomers of which the RNA is formed, together with the replicase enzyme extracted from a virus. Some original RNA molecules were added as inoculum to a tube containing the free enzyme, and free nucleotide monomers were fed continuously. After a few hours, it was observed that many copies of the original RNA molecules were formed, but subsequently, a mutant RNA appeared, much shorter because it was composed of 220 nucleotides, and prevailed over the others. The experiments performed by Eigen and his coworkers may be considered complementary to that performed by Spiegelman, since the nucleotides and the replicase were introduced into the initial mixture in the absence of RNA. A drop of the resulting solution was added to a subsequent tube with the same ingredients, and so on, until the formation of a winning molecule consisting of 220 nucleotides. The above experiences are of undeniable importance, however, they do not offer a contribution to the knowledge of what was happening in the primordial soup, because the replication of the RNA molecules is due to the presence of an enzyme whose chemical structure, and hence its informational content, is much higher than that of the molecules involved in polymerization.

Before concluding, it is important to point out that the model introduced by Eigen and Shuster represents an important tool in the description of Darwinian dynamics. It is widely employed in theoretical biology to describe the behaviour of a large population. In this approach, Eq. (10.1) is called the quasispecies equation, because it refers to an ensemble of similar genomic sequences, generated by a mutation-selection process.

10.3.3 The Chicken or the Egg?

In order to proceed, it must be realized that the uncertainties as to the priority to be attributed to proteins with respect to nucleic acids, or vice versa, reflect a real difficulty in determining which of the two kinds of component played an indispensable, and thus primary, role in the origin of life. Indeed, these two major classes of bio-polymer depend upon each other, because, according to the "Central Dogma of Molecular Biology", the flux of information proceeds in the direction going from DNA to proteins, since it is required to control their formation. In the meantime, there is a feedback moving from right to left, connected to the catalytic function exercised by proteins in the synthesis of nucleotidic molecules.

It is the question of the popular dilemma as to "which came first: the chicken or the egg?"

In order to get out of such a frustrating circularity, a self-consistent mechanism can be adopted by assuming that both of the processes take place through a network of complex chemical paths until the emergence of a symbiotic interaction able to favour evolution towards the modern genetic apparatus. For instance, Graham Cairn Smith, an English organic chemist, published a book in 1985 with the significant title: *"Genetic Takeover and the Mineral Origin of Life"*. In it, it is assumed that the prime organisms developed thanks to the catalytic action exerted by the surface of clays, consisting mainly of silicon and aluminum, together with other metals such as magnesium, and with a typical lamellar structure formed of superimposed atomic layers. All that through a series of complex chemical reactions whose occurrence is due to the ability of the clay catalytic centers to produce copies of organic products subject to modifications that are transmitted to new molecules. Then, a process of natural selection emerges, because the clay crystals catch certain molecules in the interspace between their surfaces by enhancing their replication potential. On the whole, a complex chemistry emerges, at least complex enough to evoke the presence of a prebiotic nightmare for the chemists. Actually, the concept that life is just complicated chemistry could be a way to distract attention from the real problem. For this reason, the hypothesis of the "primeval soup" began to be questioned, because the absence of supporting evidence seems to become evidence of its absence. In fact, it is reminiscent of the bootstrap phenomenon invoked in Rudolf Erich Raspe's story on *"The Surprising Adventures of Baron Munchausen"*. Curiously, something similar was invoked in the 1960s and 1970s, in a study of the world of eternally more copious subnuclear particles, by assuming that none of them existed independently. In the advocation of the concept of a "nuclear democracy", the idea that some particles were more elementary than others was discarded by orienting the researches to collect as much information as possible about their interactions, described by a mathematical tool called the S-Matrix, obtained through the quantum mechanical description of the collisions between the particles. After a few years, the approach was left to orienting the researches towards the use of increasingly powerful, and expensive, particle accelerators, in which, through the breaking of the existing particles, the subnuclear Matryoshka is enriched. This cannot occur in the prebiotic soup, because the atoms of the chemistry must be considered to be indivisible, as sanctioned by Democritus.

10.3.4   RNA World

In the second half of the last century, the idea arose that the development of life on our planet was connected to a molecular evolutionary process based on RNA. This concept was first proposed in 1962 by Alexander Rich, followed by Carl Woese in 1967, and supported by Francis Crick and Leslie Orgell, a gifted theoretical chemist who converted to the study of biology. It was baptised "RNA World" in a paper published in 1986 in Nature written by the Nobel laureate Walter Gilbert, to characterize a hypothetical phase of the evolutionary story on Earth, in which self-replicating RNA molecules proliferated before the evolution of DNA and proteins.

Significant support came in 1989, when the Nobel Prize for Chemistry was awarded to Sidney Altman and Thomas Cech for highlighting that the RNA molecules can catalyze some reactions, particularly those related to the formation of peptide bonds from which the proteinic skeleton originates. This fundamental result associated with RNA's peculiar nature of transferring genetic information suggested that it could be the basic molecule for escaping from the afore-mentioned prebiotic chemist's nightmare, because it can simultaneously play the role of the egg, as a store of information, and of the chicken, able to implement its contents. Therefore, in order to support the role of the afore-mentioned RNA World, emphasis was put on an RNA molecule (A) able to catalyze its own replication by reacting with a nucleotide 5'-triphosphate molecule (B). The process occurs through the formation of a chemical bond associated with the release of a single pyrophosphate group $(P_2O_7^{4-})$ from B through a hydrolysis reaction. The "RNA World" hypotheses include the basic assumptions that at some time in the evolution of life, genetic continuity was assured by the replication of RNA, occurring through the familiar base-pairing process. Moreover, the genetically encoded proteins were not involved as catalysts. A bit of evidence indicating that an RNA World did indeed exist on the young earth comes from the structure of the ribosome, the familiar protein factory, because the active site for peptide-bond formation lies within the core of RNA. So, as suggested by Crick, "*the primitive ribosome could have been made entirely of RNA*". Despite this, the assumption that all of the components of RNA were available in some prebiotic pool, and that they could have assembled into replicating, evolving polynucleotides without the prior existence of any evolved macromolecules, is still under discussion.

Anyway, it is interesting to mention that in the wake of RNA World, the Nobel Prize winner Jack Szostak took an important step forward through the creation of a prebiotically plausible protocell, with a fatty acid vesicle as a simple form of a cell membrane. In it, the RNA replication occurs autonomously without the contribution of enzymes. The approach is consistent with the idea that a simple primitive cell would consist of a membrane that defines a compartment containing a polymer that allows for the replication and inheritance of functional information. The deepening of the characteristics of the fatty-acid membranes has also shed light on the growth and division of the protocells, as well as on the exchange of matter with the environment.

10.3.5   *The Phase Transition Approach to the Emergence of Life*

The analysis previously developed highlights the fact that the interconnected system, including the DNA, RNA and proteins present in the cells, is too complex to have emerged in a single occurrence. It follows that in order to understand the chemical processes involved in the origin of life, attention must be focused on a relatively simple, but versatile, biopolymer that could have been present in the early biotic world. In the RNA World hypothesis, such a role is played by autocatalytic RNA sequences, so that it would be worthwhile investigating how the formation of such a functional molecule occurred through non-living chemistry. The problem can be dealt with in a general way, apt to be applied to other kinds of replicating molecular system that emerged, for instance, prior to the RNA world, or as alternatives to it.

Autocatalysis occurs when a given component A can make a copy of itself by using an available monomer E, according to the reaction

$$A + E \rightarrow 2\,A + W$$

where W is a byproduct. If the rate of the reaction is assumed to be proportional to the product of the concentrations of A and E, an exponential growth process takes place, because the concentration of A increases. Of course, this is valid only for restricted conditions, because the spatial constraints will slow the rate. Actually, our interest is focused on the case in which successive reactions, resulting from the presence of an autocatalytic effect, take place. The process can bring about the emergence of self-organized polymeric molecules whose complexity is associated with their intrinsic structure and the increase in their catalytic capacity. Of course, the intriguing point concerns the possible achievement of a bi-stable situation characterized by the dynamical presence of both living and non-living states or phases. The transition from the former to the latter creates renewed forms of organization of the chemical structures and functionalities, through which life can blossom and develop by evolution. Actually, if the living state is also present in the real world, its origin is a difficult and rare event, because spontaneously replicating systems popping up in test tubes were never found. At any rate, the previous deepening suggests approaching the problem of the origin of life by involving the occurrence of phase transition processes initiated by the fluctuations of the concentrations of a small number of molecule. Their characteristics are borrowed from thermodynamics, but moving from equilibrium systems to the more complex non-equilibrium ones in which chemical reactions are taking place.

Let us now consider a well-mixed chemical reacting system. The balances of the different types of molecule present are expressed by means of a system of ordinary differential equations, describing the evolution of their concentrations over time. Each of them is obtained by applying Eq. (2.7), in which the rates of transformation of the involved species are expressed through the product of their concentrations C_i (moles over volume) times a rate constant. Attention is focused on the evolution of

RNA polymeric molecules in a system where the precursor "food" molecules, F_1 and F_2, are supposed to be available at concentrations C_{F1} and C_{F2}. Let us assume, following Meng Wu and Paul Higgs, that the monomers, denoted by A, with concentration C_A, are synthesized from F1, with rate constant s, and can then react with F2 to produce activated monomers, A*, with rate constant a. RNA polymers An of length n are formed through the reaction of the activated species with A_{n-1}, which extends its length by one unit through a reaction with rate constant r. All molecules can be removed from the system at a rate expressed through the product of their concentration times a volumetric fluid flux Φ. The following system of ordinary non-linear differential equations is obtained describing the evolution of the mono-mer, of the polymeric chains and of the activated monomer, respectively:

$$\frac{dC_A}{dt} = sC_{F_1} - aC_{F_2}C_A - rC_AC_{A*} - \Phi C_{A_n}$$

$$\frac{dC_{A_n}}{dt} = rC_{A*}(C_{A_{n-1}} - C_{A_n}) - \Phi C_{A_n} \qquad (10.2)$$

$$\frac{dC_A^*}{dt} = aC_{F_2}C_A - rC_{A*}(C_A + C_P) - \Phi C_{A*},$$

where

$$C_P = \sum_{n \geq m} C_{A_n} \qquad (10.3)$$

where m is the length of the polymer chain at which its catalytic action for the polymerization starts. The polymerization rate constant is expressed as follows:

$$r = r_0 + kP_m, \qquad (10.4)$$

r_0 being the rate of the spontaneous polymerization process, while the second term on the right-hand side accounts for a feedback effect due to the increase in the catalytic efficiency, because it depends on the length of the polymer chain. Then, the previous equation accounts for the presence of a feedback process, due to the increase in the polymerization process as a consequence of the increase in the amount of the polymer. In order to investigate the behaviour of species concentration present in the stationary state of the system, the derivatives on the left-hand side must be put as equal to zero, as elucidated in Box 2.2. The outcome of the system of non-linear algebraic equations so obtained can be investigated by means of the afore-mentioned numerical methods. The obtained results, according to the values of the parameters k and r_0 of Eq. (10.4), define a phase diagram in which the zones of existence of the different states of the polymers can be identified. The most inter-esting aspect concerns the presence of a dead, or non-living, state, when Pm is very small and $r \approx r_0$, so that the polymerization occurs at the spontaneous rate, and of a living state, when $kPm \gg r_0$, so that the polymerization becomes autocatalytic and takes place at a much higher rate than the one corresponding to the spontaneous

Fig. 10.3 Some reactive
particles tend to form a
random distribution at
equilibrium, but if a single
particle (yellow) is subjected
to a periodic external force,
new structures appear in the
steady-state. *J. England et al
PRL 119, 038001 (2017)*

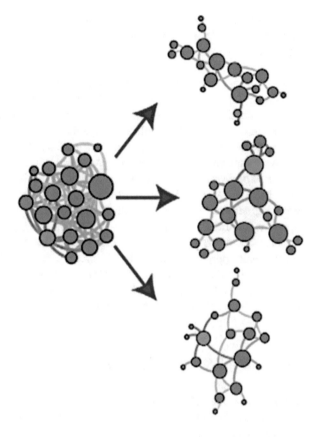

polymerization. The region of interest, of course, is the zone where both states are stable, which occurs if r_0 is fairly small and k is fairly large. If k is too small, only the dead states are stable, and if it is too large, only the living states are stable. It is important to point out that the previous result is similar to the one obtained with the model introduced by Dyson in his research on the origin of life, illustrated in Fig. 10.3. In such a case, the model has been introduced by referring to the earlier hypothesis of proteins, but since no specific equation concerning the chemical transformations were involved, it can also be invoked for the analysis of the states present in an autocatalytic process concerning the evolution towards the RNA World. Nevertheless, the presence of living and dead stable states and of possible stochastic transitions between them is also contemplated in Dyson's model. In conclusion, in the described framework, the origin of life is put forward as a transition between stable states, driven by stochastic fluctuations of the reactant concentrations. For instance, in order to model the system, a relatively small number of molecules in a localized region of space has been considered by describing their movements on a two-dimensional lattice where the reactions occur locally on single sites and diffusion takes place through the hopping of molecules to neighbouring sites. The transition to life occurs in one local region and then spreads across the rest of the surface. On the whole, it comes out that the origin of life is a rare stochastic event that is localized in one region of space due to the limited rate of diffusion of the molecules involved and that the subsequent spread across the surface is deterministic.

The concept of phase transition has been enlarged by Eric Smith and Harold J. Morowitz in order to offer a paradigmatic description of the origin of life and evolution of the biosphere through a hierarchical succession of phases associated with an increasing level of molecular organisation. Accordingly, the emergence of life should be understood as a cascade of dynamical phase transitions rearranged into structures that favor the flow of energy according to the second law of thermodynamics. Moreover, even if this approach is attractive, it is not altogether compatible with the occurrence of historical contingencies, the role of which is pervasive in biological processes, Starting from the occurrence of causal mutations, which represent important steps in the neo-Darwinism paradigm. Actually, awareness of the relevance of these historical contingencies makes biology incompatible with the deterministic approach from which the conceptualization of our scientific knowledge has matured. However, the role of history is becoming pervasive, even in events of the physical and chemical sciences, since it was realized that the presence of non-linear behaviours makes them extremely sensitive to their initial conditions.

10.3.6   Back to Thermodynamics

Nothing, including life, can violate the second law of thermodynamics, as has been demonstrated in Chap. 1. Nevertheless, luckily, life exists despite the puzzle of justifying its presence on Earth.

Progress on the thermodynamical behaviour of open systems out of equilibrium, driven by external energy sources, has been slow, all the more so because their description was always considered a hard nut to crack. Nevertheless, the situation seemed improved at the end of the last century, thanks to the work of Chris Jarzynski and Gavin Crooks, who were able to show that the entropy produced in a thermodynamic transformation can be expressed by the ratio between the probability of undergoing the process divided by the probability of undergoing the reverse process. This result was expressed through a simple but rigorous formula that, in principle, can be applied to any thermodynamic process, no matter how fast or far from equilibrium. Through its application, Jeremy England, an English scientist teaching and working at MIT, introduced an approach to the description of the transition from an a-bio- to a bio-world.

Let us focus attention on a system that exists in a set of microscopic states, indicated by i, coupled to a heat bath and obeying stochastic dynamics, similar to that introduced in Sect. 1.4, to describe the Brownian motions, occurring through a set of transitions, having probabilities $\pi(\rightarrow i|j)$, starting from a microstate i, with energy ε_i, and ending in the microstate j, with energy ε_j. Such transitions capture the essential microscopic relationships between heat and irreversibility, because the more likely a trajectory is to be forward than it is to be in reverse, the more the overall entropy is increased through the depletion of heat into the surrounding bath. But what happens throughout the entire system, which, being in some macroscopic state **I**, is observed after a time interval τ? If our attention is focused on a biological system, and τ is on the order of the time of cellular fission, the expected final state **II** could be the presence of two bacteria floating in the media instead of one, while various surrounding atoms are rearranged into new molecular combinations. The labels **I** and **II** summarize the characteristics of the experimental macroscopic state of the system, including the values of temperature, pressure, chemical composition and structure. Such information depends on the interactions with the environment, including the driving action on the transformations. Within this framework, the following generalized expression of the second law of thermodynamics is derived, which accounts for the behaviour of systems driven by an external energy source and dumping heat into a surrounding bath:

$$\beta \langle \Delta Q \rangle + \ln \left[\pi(\rightarrow \mathbf{I}|\mathbf{II}) \right] + \Delta S_{\text{int}} \geq 0. \tag{10.5}$$

In it, $\pi(\rightarrow \mathbf{I}|\mathbf{II})$ is the mentioned probability of the transition in the time τ between two macroscopic states; in principle, this can be evaluated from the transition probabilities between the microscopic states of the system.

$\Delta S_{\text{int}} \equiv S_{\mathbf{II}} - S_{\mathbf{I}}$ instead expresses the associated internal entropy change, involving any kind of transformation. It is evaluated by means of the Shannon equation (2.8), which gives us the variation of information by going from the initial to the final state of the transformation. Finally, ΔQ is the heat released into the bath and, being $\beta = 1/k_B T$, $\beta \langle \Delta Q \rangle$ is the entropy discharged into the bath.

The previous equation looks cryptic, but, in essence, states the connection between the probability of the transition from one state to another, having different informational content, with the entropy discharged into the environment. In particular, it highlights the fact that in a transformation, the more likely evolutionary outcomes, having a high value of transition probability, are the ones able to dissipate more of the energy coming from the driving actions initiated by the environment. Then, it offers a tool for evaluating the minimum theoretical amount of energy dissipation involved in apparently unusual transformations, such as the self-replication of biopolymers. Moreover, it allows for evaluation of the possible role played by some peculiar chemical reactions in evolutionary processes, including the origin of life.

Potentially, the approach can been applied to a wide range of non-equilibrium scenarios by opening a research program aimed at clarifying the relationship between the likelihood of a system to adopt a particular microscopic configuration and the amount of energy absorbed and dissipated during the system's dynamical history. In a reasonable approximation, it is assumed that, for a short period of time, the probability of the transitions between the two macroscopic states considered could be evaluated by making use of the parameters that characterize their rates, obtained through experimental information. One example is the self-replication of the RNA molecule, which, as previously illustrated, occurs through a bond-making process and a hydrolysis reaction. Following the afore-mentioned approximation, a value of ΔQ is obtained by means of Eq. (10.2), close to the experimental value of the enthalpy change of the reaction, by justifying its spontaneous occurrence. The same kind of analysis performed on DNA, which, in aqueous solution, is much more kinetically stable against hydrolysis than RNA, gives us a value of ΔQ, which exceeds the estimated enthalpy of the reaction, thus confirming that the reaction does not spontaneously occur. On the whole, it comes out that RNA is a cheap material whose emergence, perhaps, could be identified with an evolutionary take-over. Actually, for a confident application of the theory, an extensive investigation is required on the behaviour of specific systems with a chemical composition that can undergo the transitions of interest in biological systems. At the molecular level, evolution is a casual process, so that computational researches have been conducted on the stochastic behaviour of collections of reacting particles by subjecting them to periodic forcing, in order to reveal the emergence of adaptive resonances through which the particles could dissipate more energy. For instance, as illustrated in Fig. 10.3, in the absence of drive, some reactive particles tend to form a random graph at equilibrium, but if a single particle (yellow) is subjected to an oscillatory external force, the rates of formation and breakage of certain bonds are modified, and new structures appear in the steady-state. On the whole, it has been confirmed that a chemical mixture open to the exchanges of matter and energy with the environment may exhibit steady-state chemical concentrations, as was anticipated in Sect. 2.3. In biology, the existence of the cell structures is forced by fluxes of energy and matter, and so the present approach stresses the role of the energy dissipation in the emergence of evolutionary paths able to foster sinergetic interactions between distinct catalytic processes. From this perspective, the approach is reminiscent of the proposal by Iliya Prigogine in regard to the so-called "dissipative structures".

Furthermore, the intriguing hypothesis has been advanced that, besides biological self-replication, the formation of greater structural organizations can also be a means by which strongly driven systems ramp up their ability to dissipate energy. England argues that under certain conditions, matter spontaneously self-organizes itself. This tendency could account both for the internal order of living things and for the formation of many inanimate structures, such as sand dunes and turbulent vortices, present in many-particle systems driven by some dissipative process. Such an idea will likely face close scrutiny in the coming years.

10.3.7   The Revenge of Energy

"But what about energy?" asks Nick Lane, in a recent challenging book with the significant title "The Vital Question", observing that the formation of the organic compounds involved in the formation of the prebiotic macromolecules and their aggregation in cells need energy. No problem, there is plenty of it coming from outer space, mostly from the sun. In Chap. 3, we realized that the Universe has usurped from the earth the prerogative to give hospitality to complex chemical compounds, whose formation takes advantage of the energy sources present in the cosmic nebulae. The enthusiasm for these discoveries promoted many researches aimed at identifying the mechanisms by which such a cosmic chemistry began and developed, as was described in Sect. 3.4. Moreover, the presence of molecules of prebiotic interest triggered speculation and fantasies about the existence of extraterrestrial life. Thus, why worry that something similar should not also happen on Earth? Whenever energy is required for the occurrence of the physico-chemical processes involved in biological systems, the intervention of the ATP and NADH in supplying the required free energy is invoked, the former thanks to its versatility summarized in Fig. 2.3, by demonstrating that it is the fundamental magical fuel employed in most biological events, including the ones mentioned at the beginning of the present book when molecular motors were introduced. The intervention of ATP allowed to escape from the embarrassment of invoking a devil to justify the ordered movements of the kinesin and other bio-machines, whose behaviour appears to be contrary to the second law of thermodynamics. The amount of ATP involved in biological systems is enormous, for instance, a cell of 3000 μm^3 produces 10^9 molecules of ATP per second, and the details of some mechanisms that occur have been described in the previous chapters. The energy, coming from solar radiation thanks to the action of photosynthesis, is captured and stored in carbohydrates, whose molecules are thus the free energy bearers to all of the living world through the food chain. And beyond, since the fossil fuels that still account for the largest energy resource used by men originate from the organic matter buried underground millions of years ago. But, as illustrated in Chap. 6, such an excellent job is carried out by a complicated biomachine, whose emergence in the frame of evolutionary bricolage took place about one billion years after the supposed emergence of life on the planet. Thus, in order to overcome such a bottleneck, it can be hypothesized that the solar electro-magnetic radiation may have promoted certain photo-chemical processes that led to the formation of the molecules able to trigger a prebiotic chemistry. For example, in an atmosphere containing nitrogen and methane, it can be assumed that UV radiation promoted the formation of hydrogen cyanide and its derivatives. The chemist John Sutherland and his team, working at the Laboratory of Molecular Biology in Cambridge, have been able, starting from hydrogen cyanide, hydrogen sulfide and ultraviolet light from the sun, to map a set of reactions producing sugars, amino acids, ribonucleotides and glycerol. The list includes the fuel for metabolism, the building blocks for proteins, ribonucleic acid (RNA) molecules and the fats that form cell membranes. These are remarkable results, even if some doubts have been raised as to the possibility that an atmosphere containing hydrogen cyanide in the presence of UV radiations could be considered a plausible primordial condition. After all, no known form of life uses cyanide, while it is known that UV radiation exerts a detrimental effect on organic molecules, rather than promoting their formation. For these reasons, the interest in re-examining the chemical reactions involved in prebiotic processes in the spirit of metabolic philosophy was reconsidered. From this perspective, Gunter Wachtershauser, a chemist and lawyer working at the Munich Patent Office, starting from 1980, developed a model of the occurrence of an evolutionary high-temperature metabolic chemistry, invoking a scheme of autocatalytic reactions whose occurrence is due to the reduction of some sulfur compounds, such as H_2S and FeS, transformed into pyrite, with the production of energy:

$$FeS + H_2S \rightarrow FeS_2 + H_2 + energy.$$

This process takes place near black smokers, which are submarine hydrothermal oceanic sources out of which mineral rich and high-temperature water flows. Thanks to such an energetic boost, hydrogen reacts with the carbon dioxide in seawater to synthesise organic molecules starting from formic acid HCOOH. A cascade of coupled reactions follows, building up the most complex organic molecules involved in cell formation. Therefore, according to the proposal by Wachtershauser, life originated on mineral surfaces in liquid bubbles as cells in which metabolism predates genetics, with iron sulfides as energy sources, through an autocatalytic process towards a self-replicating cycle of reactions. It is a conjectured cascade of coupled reactions that build up the most complex organic molecules purported to be involved in cell formation.

Despite the reasonableness of some of the afore-mentioned hypothesis, the origin of life looks appears to be an enigma. In reality, in recent years, quite a bit of evidence about certain characteristics of the processes involved has emerged, thanks to researches on molecular biology, biochemistry, genomics and geology. In partic-ular, relevant results have been obtained thanks to the structural investigations performed by means of X-ray chrystallography, as has been illustrated in the previous chapters, particularly for that which concerns the amazing characteristics of biochemical machineries. Their operation is determined by the employment of energy, which becomes the most relevant ingredient of life, additionally because without energy, the same information is useless. A fascinating example is offered by the remarkably large protein complexes that catalyse the cascade of reactions involved in the energy transmission that occurs in mitochondria, as illustrated in Sect. 4.4. Also quite interesting are the clues as to the existence of metal active centers, such as for instance, iron sulphide, in some enzymes.

In cells, the source of energy is molecules with a high content of free energy, such as carbohydrates, which are exchanged through metabolic networks of chemical reactions, typically glycolysis, followed by the Krebs cycle presented in Sect. 4.4. Therefore, it appears reasonable to focus attention once again on the role of metab-olism in the origin of life.

Nick Lane, a researcher in evolutionary biochemistry working at the University College of London, who wrote the previously-mentioned book, agrees with the proposal that the most plausible location at which life might have begun is the alkaline hydrothermal vents. Specifically, he places attention on the one near the Mid-Atlantic Ridge, on the deep ocean floor discovered only in 2000, after being predicted by the pioneering geochemist Mike Russell. It has the right traits, partic-ularly the presence of masses of warm energetic minerals pouring out to form calcium carbonate chimneys full of micropores.

This means that, by mutating the language of chemical reaction engineering, the hydrothermal vents can be considered the flow reactors of life. Starting from the primitive world, their micropores would have contained and concentrated the ingredients required to create the organic chemical precursors of life, while the hot water emitted, rich in iron and sulphur, would have created temperature gradients. The biochemical mechanisms and structures that evolved from the described systems are the same as those present in the mitochondria that are our power house.

In Chap. 4, it was illustrated how the phosphorylation of ADP to ATP is associated with the transport of protons, and that such a phosphorylation reaction occurs thanks to the action of a "splendid molecular machine", the ATP synthesis. This coupling, known as Chemiosmosis, is universal, because almost all cells harness electrochemical proton gradients across membranes to drive ATP synthesis, powering the biochemical processes that occur in living organisms. This finding suggests that it arose very early in evolution, even if its origin is still obscure. The quite interesting point now is that, in the afore-mentioned alkaline hydrothermal systems, natural proton gradients are present across the thin inorganic barriers of interconnected micropores within the deep-sea vents, suggesting a possible abiotic origin of chemiosmosis. The important question that arises is whether the afore-mentioned aspects can offer suggestions as to what the driving forces that impelled the origin of life were.

A simple model of a cell energetically fed by a natural proton gradient is illustrated in Fig. 10.4. Enclosed by a semi-permeable membrane, it sits at the interface between alkaline and oceanic acidic fluid streams. The fluids inside are continuously replaced. In fact, the hydroxide ions (OH^-) can flow into the cell from the alkaline side at the bottom by simple diffusion across the membrane, while protons (H^+) enter in a similar manner from the acidic side from the top. Inside the protocell, H^+ and OH^- can neutralize into water, or leave from either side, so that the internal level of acidity depends on the water dissociation equilibrium and the relative influxes of each ion. A protein capable of exploiting the natural proton gradient sits on the acidic side, allowing energy assimilation via ATP production. The afore-mentioned hypothesis implies that such electrochemical ion gradients across membranes, able to drive metabolism, could be conserved as genetic code. The hypothesis also suggests that ion gradients arose at the beginning of evolution, so that they may have played a role in the origin of life. The inside of the protocell is relatively alkaline, so that the driving force for performing work through the afore-mentioned protein on the "acid ocean" side is therefore significantly greater than the work performed on the "alkaline" side. Such a situation, which can enable ATP synthesis, may be the starting point of an evolutionary process towards the present structure of mitochondria. In conclusion, the hypothesis that some form of chemiosmotic coupling probably evolved very early in the history of life, arguably before LUCA, is intriguing and stimulating for further researches, all the more so because it leaves open the question as to how, and why, it happened.

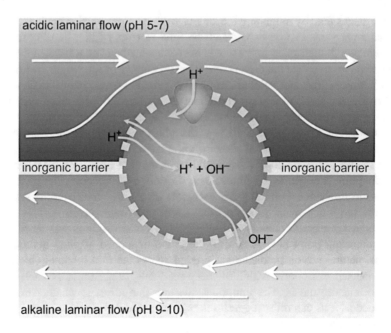

acidic laminar flow (pH 5-7)

inorganic barrier $H^+ + OH^-$ inorganic barrier

alkaline laminar flow (pH 9-10)

Fig. 10.4 A cell enclosed by a semi-permeable membrane, which sits at the interface between alkaline and oceanic acidic fluid streams, is powered by a natural proton gradient. Hydroxide ions can flow into the cell from the alkaline side by simple diffusion with protons entering in a similar manner from the acidic side. *Victor Sojo et al., PLOS BIOLOGY, August 2014 | Volume 12 | Issue 8 | e1001926*

10.3.8   A Merging Point?

The preceding analysis, based on the assumption that metabolism is separated from replication, is afflicted by the difficulty of justifying the convergence of the two aspects. Not a simple task, because the involved chemistry refers to different molecules, respectively, proteins for metabolism and the nucleic acids for replication, which have complementary functionalities. The idea that such a convergence has occurred in an extraordinary single sinergic event, as suggested in Fig. 10.4, seems unlikely, while it is reasonable that in the long period of time leading to the emergence of LUCA, the synthesis of the afore-mentioned specific molecules, proteins and DNA happened separately. For example, the former process could have occurred through the approach suggested by Dyson, involving a statistical inheritance, while the latter was preceded by the comparison of the RNA World, which later resulted in the formation of the nucleotide polymers, according to the kinetic model suggested by Eigen. Within this framework, the challenge must be faced of constructing a realistic picture of the origin of the RNA World, including the details, still obscure, as pointed out by by M.P. Robertson and G.F. Joyce, as to the logical order of events, beginning with prebiotic chemistry and ending with DNA/protein-based life. At any rate, the emergence of such a presumed RNA World should be viewed as a milestone in the early history of life on Earth, and further progress will depend on new experimental results addressed to face the problems concerning the RNA-based cellular processes. Anyway, the first life form capable of reproducing was already a very complex system capable of self-maintenance. It was a natural technological wonder composed of interacting parts, involving both their material characteristics and the functions they were able to exercise. The further phase of the process that brought about the formation of a cell started when the two systems began to collaborate. It is certainly the most complex and most mysterious event, but also the most interesting, because only when the RNA World and the metabolic world merged did the formation of modern cells begin. This phase required the intervention of the familiar molecular machines that played an important role in the development of cellular complexity and functionality, by involving energetic aspects, through the intervention of the ATP produced in mitochondria, and structural aspects, through the production of proteinic molecules by ribosomes. With regard to the first aspect, Nick Lane's recent proposal as to the origin of chemiosmotic processes offers an important outlet, since it occurs through the intervention of the electrochemical energy associated with the presence of potential gradients in the alkaline hydrothermal vent, more suitable in the early stages of molecular evolution than the employment of other forms of energy, including solar. As to the ribosome, it is a complex molecular machine that can have as many as 80 proteins interacting with multiple RNA molecules, so it makes sense that its assemblage is the result of a long and complicated process of gradual co-evolution typical, as described, of the Darwinian theory. Finally, it is important to mention that the modern ribosome was largely formed at the time of LUCA, and thus, as previously mentioned, its earliest origins likely lie in the RNA World, stressing once again the importance of such a phase of biomolecular evolution.

10.4   The First Synthetic Cell

The difficulties advanced in the preceding chapters to express a definition of life can be summarized in the following sentence by Gerald Joyce:

> If there is a sine qua non of life, it is the ability to undergo Darwinian evolution, and to have history in molecules. Chemistry does not have history, while biology has history. The dawn of life is the dawn of biological history written in the genetic molecules that are carved through Darwinian processes.

The previous statement focuses interest on the characteristics of genes, in order to investigate the role of their complexity in the tendency to evolve towards cells, through differentiation and specialization as a consequence of the increase in their structural variety. The experimental researches confirm that the genomic sequences from diverse phylogenetic lineages are associated with notable increases in genome complexity, moving from prokaryotes to multicellular eukaryotes. Therefore, Craig Venter and his coworkers oriented their efforts towards creating a "hypothetical

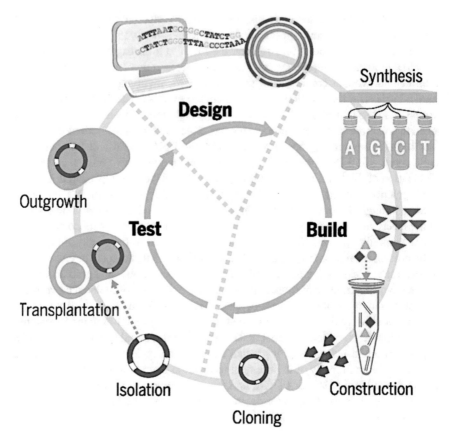

Fig. 10.5 The approach to the synthesis of an artificial cell. It follows a circular procedure in which synthesis and computing were interactively involved

minimal genome". In March 2016, they published, in Science, a paper emblematically entitled *"Design and synthesis of a minimum bacterial genome"*. The working approach, illustrated in Fig. 10.5, followed a circular procedure in which synthesis and computing were interactively involved. It resulted that they were successful in creating a genome containing about half a million base pairs organized into 473 genes, transplanted into cells of tiny and fast replicating bacteria, *Mycoplasma mycoides* and *Mycoplasma capricolum*, respectively. Called JCVI-syn3.0, "or syn3.0" for short, it was, and is yet, the smallest obtained synthetic genome, thus a good result for research into minimal cellular genomes. In fact, it retains almost all of the genes required for the synthesis and processing of macromolecules, together with about 150 genes, suggesting the presence of yet undiscovered functions essential for life. In conclusion, Syn3.0 turned out to be a versatile platform for investigating the core functions of life and for inspiring future synthetic biology efforts in the introduction of novel production pathways, following the approach that is at the core of metabolic engineering. In fact, as described in the previous chapter, the aim is to create novel biological functions and systems not present in nature by combining biology with chemistry, information science and engineering. In this approach, microbial organisms are utilized as cell factories for the bioconversion of renewable resources into bulk or high value chemicals. The introduction of novel production pathways is, in fact, at the center of metabolic engineering. If DNA, particularly the genome, is the software of life because all of the characteristics and functions of living cells are written into the genetic code, biologists can now act as software engineers, able to rewrite biological operating systems. If the desired sequence does not exist in nature, a synthetic approach must be followed. The vision in synthetic biology is to have the capacity to computer-design and build a DNA that produces a biological cell with a predictable outcome. Significant advances are in progress in DNA design at the gene and pathway level, but, despite the improvements in genomics and synthetic biology, much of the path has yet to be covered for understanding the functions of all of the genes present in a synthetic cell. The minimal cellular systems will become the test for understanding the rules governing genome design. The efforts are intriguing and stimulating, even if obtainment of a minimal cell represents only the start of a fascinating adventure in progress, addressed at inspiring a turning point for the progress of human society in its cultural, technological and social aspects.

References

Dyson, Freeman. *Origins of Life*, Cambridge University Press, 2004. Smith, John Maynard, Eors Szathmary. *The major Transition in Evolution*, W.H.FREEMAN Ptectrum, Oxford, 1995.

Davies, Walke, PCW. 2013. *The algorithmic origins of life* J. R. Soc Interface 10: 20120869. https://doi.org/10.1098/rsif.2012.0869.

Robertson Michael P, Gerald F. Joyce. *The Origins of the RNA World*, Cold Spring Harb Perspect Biol 2012;4:a003608.

Joyce Gerald. *Bit by Bit: The Darwinian Basis of Life*, May 2012 I Volume 10 I Issue 5 I e1001323.

Meng Wu, Paul G Higgs. *The origin of life is a spatially localized stochastic transition*, Biology Direct 2012, 7:42 http://www.biology-direct.com/content/7/1/42 Perunov Nikolay, Robert A. Marsland, and Jeremy L. England. *Statistical Physics of Adaptation*, PHYSICAL REVIEW X 6, 021036 (2016).

Venter J. Craig. *Life at the Speed of Life*, Viking, 2013.

Gibson Daniel. *Programming biological operating systems: genome design, assembly and activation*, NATURE METHODS I VOL.11 NO.5 I May 2014.

Mann Stephen. *Systems of Creation: The Emergence of Life from Nonliving Matter*, ACCOUNTS OF CHEMICAL RESEARCH 000–000 XXXX Vol. XXX, No. XX.

Kachman Tal, Jeremy A. Owen, Jeremy L. England. *Self-Organized Resonance during Search of a Diverse Chemical Space*, PRL 119, 038001 (2017).

Szostak, Jack W., David P. Bartel & P. Luigi Luisi, *Synthetizing Life*, NATURE|VOL409|18JAN-UARY2001|www.nature.com.

Eigen Manfred. *Self-organization of matter and the evolution of biological macromolecules*, Naturwissenschaften 58 (1971) 465–523.

Woese Carl. *Interpreting the universal phylogenetic tree*, Proc. Natl. Acad. Sci. USA 97 (2000) 8392–8396.

Yockey H.P. *An application of information theory to the Central Dogma and the sequence,* J. Hypothesis Theor. Biol. 46 (1974) 369–406.

Herschy Barry, Alexandra Whicher, Eloi Camprubi, Cameron Watson, Lewis Dartnell, John Ward, Julian R. G. Evans, Nick Lane. *An Origin-of-Life Reactor to Simulate Alkaline Hydrothermal Vents,* J Mol Evol (2014) 79:213–227 DOI 10.1007/s00239-014-9658-4.

Donald E. Johnson. *What Might Be a Protocell's Minimal "Genome?", The First Gene,* David L. Abel, Editor, 2011, pp 287-304.

Smith Eric, Harold J. Morowitz. *The Orign and Nature of Life on Earth,* Cambridge University Press, 2016.

Clyde A. Hutchison III, Ray-Yuan Chuang, Vladimir N. Noskov, Nacyra Assad-Garcia, Thomas J. Deerinck, Mark H. Ellisman, John Gill, Krishna Kannan, Bogumil J. Karas, Li Ma, James F. Pelletier, Zhi-Qing Qi, R. Alexander Richter, Elizabeth A. Strychalski, Lijie Sun, Yo Suzuki, Billyana Tsvetanova, Kim S. Wise, Hamilton O. Smith, John I. Glass, Chuck Merryman, Daniel G. Gibson, J. Craig Venter. *Design and synthesis of a minimal bacterial genome,* Science, 25 MARCH 2016 • VOL 351 ISSUE 6280.

Rutherford Adam. *Creation,* CURRENT, New York, 2013.

Cox Brian, Andrew Cohen. *Human Universe,* William Collins, London, 2014.

Dennett, Daniel C., *From Bacteria to Bach and Back. The Evolution of Mind,* 2017.

Glossary[1]

Activation energy This is the minimum amount of energy required to activate a couple of molecules into the state from which they can undergo a chemical transformation. Even the energy-releasing, or exothermic, reactions require such an amount of energy input before they can proceed to the energy-releasing step.

Algorithm In mathematics and computer science, this refers to a procedure that solves a recurrent problem through the conduction of a sequence of specified actions. A computer program can be viewed as an elaborate algorithm. Algorithms are widely used throughout all areas in which information technology is involved.

Allostery The phenomenon through which a ligand binds to a specific receptor site on a protein, changing its shape and altering its affinity for another ligand at a second site. It follows that the catalytic function of an enzyme may be modified by interaction with small molecules acting not at the active site, but also at a spatially distinct, allosteric, site of different specificity.

Archaebacteria Class of prokaryote bacteria very different from common bacteria, and from all other modern lifeforms. Named archaebacteria, from 'archae' for 'ancient', these unique cells are thought to be modern descendants of a very ancient lineage of bacteria that evolved around sulfur-rich deep sea vents.

ATP (adenosine triphosphate) A biomolecule constituted of a nucleotide (adenine) plus three phosphate groups attached. It is the most important molecule involved in energy transport in living organisms. Its reaction with water detaches one of its phosphate groups, by giving rise to the formation of ADP (adenosine diphosphate) and phosphoric acid by liberating free energy.

---

[1]*The space reserved for the different headwords is not uniform, since in some cases, it is aimed at deepening and integrating the text, while in others, it only highlights the lexical content of some of the recurring words.*

© Springer Nature Switzerland AG 2018
S. Carrà, *Stepping Stones to Synthetic Biology*, The Frontiers Collection,
https://doi.org/10.1007/978-3-319-95459-2

Bacterium A member of a large group of unicellular microorganisms that have cell walls, but lack organelles and an organized nucleus. Bacteria are widely distributed in soil, water, and air, or in the tissues of plants and animals. Formerly included in the plant kingdom, they are now classified separately (as prokaryotes). They play a vital role in global ecology because of the chemical changes they bring about, including those of organic decay and nitrogen fixation. Much modern biochemical knowledge has been gained from the study of bacteria, as they grow easily and reproduce rapidly in laboratory cultures.

Bit Binary code, contraction of bi(nary) (digi)t. In information theory and computer science, it is the unit of measurement of the information content of a message. More precisely, a bit is the amount of information that solves the uncertainty between two alternatives (open, close, zero or one).

Boltzmann distribution law This is also called a Gibbs distribution. In statistical mechanics, it is a probability distribution, or frequency distribution, of the n_0 particles present in a system over the various possible states. The number n_k of particles present in k-th state is given by

$$n_r = n_0 e^{-\varepsilon_r / k_B T},$$

where ε_r is the energy of the state and k_B the Boltzmann constant.

Boolean algebra Introduced by George Boole in 1854, this is also called Binary Algebra or logical Algebra. It is used to analyze and simplify digital (logic) circuits, by making use of only the binary numbers, i.e., 0 and 1.

Brownian motion Expression that encompasses various physical phenomena in which some quantity is constantly undergoing small random fluctuations. It was baptized with the name of the Scottish botanist Robert Brown, the first to study such fluctuations (1827), by observing the random motion of particles suspended in a fluid, resulting from their collision with the fast-moving molecules present in it.

Carbohydrate A class, also known as *watered carbon*, that includes natural organic compounds and their derivatives. In the early part of the nineteenth century, substances such as wood, starch, and linen were found to be composed mainly of molecules containing atoms of carbon (C), hydrogen (H), and oxygen (O), having the general formula $C_x(H_2O)_y$, commonly used to represent many carbohydrates.

Catalysis This indicates the increase of the rate of a chemical reaction due to the presence of a substance not consumed during said reaction. Such rates depend upon different factors, including the chemical nature of the reacting species and the conditions to which they are exposed. Many industrial processes depend upon catalysts for their success, but, most importantly, the peculiar phenomenon of life would hardly be possible without the biological catalysts known as enzymes.

Cell Cells are the basic building blocks of all living things, and play many important roles in different vital functions. The human body is composed of trillions of cells that provide structure for the body itself, take in nutrients from food, convert those nutrients into energy, and carry out specialized functions. Moreover, cells contain hereditary material, and thus can make copies of themselves.

Central dogma The central dogma of molecular biology, first articulated by Francis Crick in 1958, concerns the flow of genetic information within a biological system by stating that "DNA makes RNA and RNA makes protein". From recent research, it has emerged that some aspects of the central dogma are not entirely accurate. Despite this, it still plays a functional role in the interpretation of cell behaviour.

Chemical energy Energy stored in the bonds of chemical compounds. It may be released during chemical reactions, called exothermic, in the form of heat. Endothermic reactions that require an input of heat to proceed may store some of that energy as chemical energy in newly formed bonds.

Chemical equilibrium This is the condition at which, in the course of a reversible chemical reaction, no change in the amounts of reactants and products occurs, because the two opposing reactions proceed at an equal rate, and hence there is no net change in the amount of the involved substances. At the equilibrium point, the reaction may be considered to be complete, because, for the specific conditions at which the reaction is performed, the maximum conversion of reactants to products has been attained.

Chemiosmosis This indicates the movement of ions across a semipermeable membrane down their electrochemical gradient. In cellular respiration and photosynthesis, it is associated with the generation of adenosine triphosphate (ATP) by adenosine diphosphate (ADP), thanks to the movement of hydrogen ions across the appropriate membrane.

Chromosome In cells, DNA is packaged into thread-like structures present in the nucleus, called chromosomes. Each is made up of a strand of DNA tightly coiled many times around some proteins, called histones, that support the structure. Chromosomes are not visible, even under a microscope, when the cell is not dividing. However, the DNA that makes up chromosomes becomes more tightly packed during cell division, becoming visible under a microscope. Each chromosome has a constriction point called the centromere, which divides the chromosome into two sections, or arms. The short one is labeled the "p arm", while the long one is labeled the "q arm." The location of the centromere on a chromosome gives rise to its characteristic shape, which can be used to describe the location of specific genes.

Codon This is the sequence of three adjacent nucleotides on a strand of nucleic acid (particularly DNA) that constitutes the genetic code for a specific amino acid that is to be added to a polypeptide chain during protein synthesis.

Complexity The peculiarity of a system composed of many interacting parts that exhibits emergent properties.

Cyanobacteria Microorganisms belonging to a phylum of bacteria that obtain their energy through photosynthesis. They are the only photosynthetic prokaryotes able to produce oxygen. The name comes from their blue color.

DNA (deoxyribonucleic acid) A helical molecule located in the cell nucleus that stores the information to make proteins and to direct the development of the cell itself.

Emergence Arising from new properties from the interaction of the many parts of a system.

Energy A word introduced by Aristotle to indicate the ability to act, but now entered into the common language because it concerns something that is able to perform a work that can be obtained from oil, coal, wind, sun and atomic nuclei. We also know that it cannot be obtained from nothing and that it is present in different forms.

Enthalpy This is an effective way to express the energy content of a thermodynamic system. It is equal to its internal energy, associated with atomic and molecular motions, plus the product of pressure P and volume V, so that it is given by $H = U + PV$. More specifically, it includes the internal energy that is required to create a system, plus the amount of energy required to make room for it by displacing its environment and establishing its volume and pressure.

Entropy Usually indicated by the letter S, this is a quantity defined for each state of equilibrium of a system; its value depends on the variables that characterize the state, particularly temperature, pressure and composition. Isolated systems with constant energy are subjected to spontaneous transformations towards the state of equilibrium corresponding to the maximum value of entropy. Entropy also allows us to characterize the quality of the different forms of energy, because it increases by decreasing its entropy content, according to the ranking:

$$\text{gravitational} > \text{nuclear} > \text{electromagnetic} > \text{thermal}$$

The Universe is subject to very active[?] transformations through which a degradation from the highest quality of energy to the lowest is in course.

Enzyme This is a macromolecular biological catalyst that accelerates biochemical reactions. Almost all metabolic processes in a cell need enzymatic catalysis in order to occur at rates fast enough to sustain life. Enzymes also have valuable industrial applications, concerning the fermentation of wine, leavening of bread, curdling of cheese, and brewing of beer, that have been practiced from the earliest eras.

Epistasis This is the study of the interactions between genes. It is of fundamental importance for understanding both the structure and the function of genetic pathways and their evolutionary dynamics.

Equilibrium In physics, this is the condition of a system at which neither its state of motion nor its internal energy change with time. According to thermodynamics, the equilibrium state of an isolated system corresponds to the maximum value of its entropy.

Eukaryotic Eukaryotic cells contain a nucleus and organelles, and are enclosed by a plasma membrane. They are larger and more complex than prokaryotic cells, which are found in Archaea and Bacteria, the other two domains of life. The organisms with eukaryotic cells include protozoa, fungi, plants and animals.

Evolution In agreement with the original meaning of the word, '*unfolding*', this is the story of the changes that occurred over time in the biological world. All that by deepening the involved paths, the implied mechanisms and the influence of the emergence of something from something else. Biological evolution, despite still being the object of discussions and researches, has reached a remarkable degree of ripeness, thanks to the contribution of genetics through its connections with information science. Moreover, its role is not confined to biological systems, but rather it touches all of human culture, by encompassing the development of technologies and the social and economical behaviour of our society.

Exergy The thermodynamic quantity that expresses the maximum work that can be obtained in recovering the equilibrium condition between a system and its environment (see Eq. 5.1).

Fatty acid This is a carboxylic acid with a long aliphatic chain, either saturated or unsaturated. Most naturally occurring fatty acids have an unbranched chain of an even number of carbon atoms, from 4 to 28. Fatty acids are usually derived from triglycerides or phospholipids.

Free energy This is the energy available to perform work. It is expressed by the difference between the total energy content of a system minus its useless part associated with the thermal motion of molecules. Our attention is focused on Gibbs free energy, expressed as $G = H - TS$. From it, the chemical potential is obtained by means of Eq. (2.11).

Gene This is a sequence of molecules of DNA, or RNA, which code for other molecules, specifically the proteins that carry out well-defined functions in a living organism. During gene expression, which is the process by which information from a gene is used in the synthesis of a gene product, the DNA is first copied into RNA, which can be directly functional or can be the intermediate template for synthetizing a protein suitable for performing the afore-mentioned function.

Genetic code This includes the set of rules by which the information encoded within genetic material is translated into proteins by living cells. Translation is accomplished by the ribosome, which links amino acids in an order specified by messenger RNA (mRNA).

Hydrogen bond This is the bond resulting from the intermolecular attraction between two polar groups that occurs when a hydrogen (H) atom, covalently bound to a highly electronegative atom, such as nitrogen (N), oxygen (O), or fluorine (F), experiences the electrostatic field of another highly electronegative atom nearby.

Hydrophilic group This is a "water-loving" group, which, when it is present in a molecule, forms hydrogen bonds, thus enhancing its solubility in water. For this reason, methanol, ethanol, n-propyl alcohol, and isopropyl alcohol are miscible with water, while alcohols with higher molecular weights tend to be less water-soluble.

Hydrophobic effect This is the tendency of non-polar substances to aggregate in aqueous solutions. Therefore, it describes their segregation from water favored in the hydrogen bonding that minimizes the area of contact between water and non-polar molecules. Hydrophobicity is responsible for the separation in mixtures of oil and water and for many effects related to biology, including the formation of vesicles and cell membranes.

Information Its mathematical definition, introduced by Claude Shannon, is based on the probabilities of available possibilities. It is closely related to the definition of entropy introduced by Willard Gibbs in his approach to statistical mechanics.

Kinesin This belongs to the class of motor proteins, such as myosin and cytoplasmic dynein, present in eukaryotic cells. They move along microtubule filaments, and are powered by the hydrolysis of adenosine triphosphate (ATP). Their active movements support several cellular functions.

Krebs cycle The Krebs cycle, or citric acid cycle, is the central driver of cellular respiration, because it includes a series of chemical reactions employed by all aerobic organisms to generate energy. Its importance to many biochemical pathways suggests that it was one of the earliest parts of cellular metabolism to evolve.The acetyl CoA,

originally derived from glucose, is its starting material, through which a series of redox reactions harvests much of its bond energy in the form of NADH and ATP molecules that are the fuels of the biochemical world.

LUCA Acronym for "Last Universal Common Ancestor", which is assumed to be the common origin of all living organisms. According to many scientists, it was not much more than an assemblage of molecules in a primordial soup from which, under the evolutionary pressure of the environment, more complex forms developed. Others scientists, instead, believe that it already had a rather complex structure, similar to a cell.

Membrane A layer formed by lipids and proteins that delimits cells and cellular organelles. These layers not only create and delimit intracellular compartments, but also provide large surfaces where the biochemical reactions necessary for vital functions can take place. The molecules of which biological membranes are composed are prevalently phospholipids, in which both hydrophobic and hydrophilic portions are present, because they possess long lipid chains, while also providing water soluble ends.

Messenger RNA This is a large family of RNA molecules that convey genetic information from DNA to the ribosome, where they specify the amino acid sequence of the protein that is produced by taking advantage of gene expression.

Metabolic engineering This is the practice of optimizing genetic and regulatory processes within cells to increase the production of certain substances. These processes involve a series of biochemical reactions and enzymes that allow cells to convert raw materials into the molecules required for different applications.

Metabolism This is the set of life-sustaining chemical transformations that occur within cells.

Microscopic reversibility Principle formulated by Richard Tolman that provides a dynamic description of the equilibrium conditions at which no net change in some given property of a physical system is observable. For instance, in a chemical reaction, no change takes place in the concentrations of reactants and products, even if a continuous activity is occurring on a microscopic (i.e., atomic or molecular) level. The principle of microscopic reversibility, when applied to a chemical reaction that proceeds in several steps, essentially states that at equilibrium, each individual reaction occurs in such a way that the forward and reverse rates are equal. In this form, it is known as the principle of detailed balance.

Microtubules Rigid hollow rods, approximately 25 nm in diameter, that are the third principal component of the cytoskeleton, which is a complex network of interlinking filaments and tubules that extend throughout the cytoplasm, from the nucleus to the plasma membrane Microtubules are dynamic structures that undergo continual assembly and disassembly within the cell. They determine cell shape and are involved in a variety of cell movements, including the intracellular transport of organelles.

Mitochondria Rod-shaped organelles that act as the power generators of the cell by converting oxygen and nutrients into adenosine triphosphate (ATP), which is the energy currency of the cell, because it powers their metabolic activities. The process is called aerobic respiration.

Molecular motors The biological molecular machines are the essential agents of movement in living organisms, being devices that convert energy into motion. Many protein-based molecular motors harness the free energy released by the hydrolysis of ATP in order to perform mechanical work. The energetic efficiency of such molecular devices can be superior to currently available man-made motors.

Morphogenesis This is the biological processes that causes an organism to develop its shape. It is one of the three fundamental aspects of developmental biology, along with the control of cell growth and cellular differentiation, that are unified in the evolutionary developmental biology called evo-devo. The shaping of an organism occurs through embryological processes of differentiation of cells and tissues, together with the development of the organ systems, according to the genetic "blueprint" of the potential organism and the environmental conditions.

Network In complex systems, the presence of interactions form *networks*, where each node interacts with only a small number of other selected nodes. Actually, networks exist everywhere and at every scale, pervading technology, the environment and social behaviour. Significant examples include the Internet, power grids, transportation systems, food webs and ecosystems. The brain is a network of nerve cells connected by axons, while cells are networks of molecules connected by biochemical reactions. Societies, too, are networks of people linked by friendship, family, and professional ties.

Non-linearity In non-linear systems, the change of the output is not proportional to the change of the input. Because most systems present in nature are inherently non-linear, non-linearity is of interest to mathematicians and engineers, along with many other scientists, including biologists. In contrast with much simpler linear systems, the dynamic behaviour of non-linear systems may appear chaotic, unpredictable, or counterintuitive. Such behavior is mathematically described by a system of differential equations that cannot be expressed as a linear combination of the variables present in them.

As non-linear dynamical equations are difficult to solve, they are often approximated by linear equations, but such an approach is effective only within some range of the input values. Unfortunately, some interesting phenomena, particularly the presence of singularities where the system blows up or becomes degenerate, are hidden by linearization. Even if many aspects of the dynamic behavior of a non-linear system can appear to be counterintuitive, unpredictable or even chaotic, they cannot be discarded, because they are often the reasons for inaccurate long-term forecasts.

Nucleotides Organic molecules that are the building blocks of nucleic acids, and thus serve as the monomer units of the two polymers, deoxyribonucleic acid (DNA) and ribonucleic acid (RNA), both being the essential biomolecules in all lifeforms on Earth.

Phase transition In physical chemistry, a phase is a region of space, throughout which all physical properties of a material, including density and chemical composition, can be considered uniform. In other words, a phase is a material region that is chemically uniform, physically distinct, and often mechanically separable. An example is offered by water, which can be present in three phases: solid (ice), liquid and gas. Each exists in a zone corresponding to the given ranges of the values of temperature T and pressure P, as illustrated in the figure. At the point of convergence of the separation lines, called triple, the three phases coexist. By varying the temperature or pressure at the crossing of a separation curve, a phase transition occurs with a sharp change in the properties of the system. Specifically, starting from the region of liquid water, it becomes vapour if we move down vertically, or ice if we move horizontally from right to left.

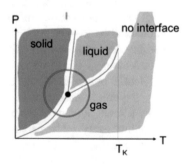

Actually, many complex systems, including not only those physics, but also those in chemistry, biology and even social groups, display different patterns, each of which corresponds to their internal organization, so that they can be compared to different phases. Such phases are usually separated by sharp boundaries whose crossing implies a deep change of the system's characteristics and behaviour. Some change may be catastrophic, and the understanding of it can throw light on the intimate characteristics of the system under examination. It follows that, of course, it is tempting also to invoke the occurrence of a phase transition in the prebiotic world for the origin of life.

Phospholipids The structure of the phospholipidic molecules generally consists of a couple of hydrophobic fatty acid tails and a hydrophilic head consisting of a phosphate group. The membrane of cells and cellular organelles are microscopically thin structures formed from two layers of phospholipid molecules. Such membranes separate individual cells from their environments by compartmentalizing the cell's interior into structures that carry out special functions.

Photosynthesis This is the process through which green plants and cyanobacteria are able to transfer sunlight energy to molecular reaction centres for its conversion into chemical energy. In it, two massive protein complexes split water and carbon dioxide and forge new energy-storing bonds present in sugar molecules by producing oxygen that is the key to planetary life. Moreover, the capture of photons and the transfer of their energy into complex molecules is at the basis of the organization of the vital processes.

Phylogenetic tree This is a diagram that represents the evolutionary relationship among organisms. The patterns of branching reflect how species or other groups evolved from common ancestors.

Polymerase chain reaction (PCR) This is a technique used to generate thousands to millions of copies of a particular DNA sequence. It was developed in 1984 by Kary Mullis, an eccentric American biochemist, who received the Nobel Prize in 1993, even though the basic principle of the technique was already described by Gobind Khorana in 1971. Mullis had the idea to use a pair of primers to bracket the desired DNA sequence, generated through a continuous doubling accomplished through specific catalytic proteins known as polymerases, able to synthesize long polymeric chains of nucleic acids. At present, PCR is a common, often indispensable, technique used for a variety of applications, ranging from the daily practicalities of medical diagnosis to those performed in courts of law.

Prokaryotic cell Bacteria and Archaea are classified as prokaryotes, while animals, plants, and fungi are made up of eukaryotic cells, which are simple, single-celled organisms that lack a nucleus and membrane-bound organelles. Prokaryotic cells are not divided on the inside by membrane walls, but consist instead of a single open space.

Protein Proteins are macro-biomolecules consisting of chains of amino acids linked to one another by a peptide bond, i.e., a link between the amino group of an amino acid and the carboxylic group of another amino acid, created through a condensation reaction through elimination of a molecule of water. Proteins perform a wide range of functions within living organisms, including the catalysis involved in metabolic processes.

Protocell (or protobiont) This is a self-organized, endogenously ordered, spherical collection of lipids proposed as a stepping-stone to the origin of life. A central question in evolution is how simple protocells first arose and how they could differ in reproductive output, thus enabling the accumulation of novel biological functions.

Quasispecies A viral quasispecies is a group of viruses related by similar mutations, competing within a highly mutagenic environment.

Ribosome This is a complex molecular machine, present within all living cells, that serves as the site for the synthesis of biological protein. In it, the amino acids are linked together in the order specified by the messenger RNA molecules.

RNA Ribonucleic acid, which is a polymeric molecule essential in various biological roles, such as coding, decoding, regulation, and the expression of genes. RNA and DNA are nucleic acids; together with lipids, proteins and carbohydrates, they constitute the four major essential macromolecules for all known forms of life.

RNA World This indicates a hypothetical stage in the evolutionary history of life on Earth. In it, it is assumed that self-replicating RNA molecules proliferated before the evolution of DNA and proteins.

Self-organization This is the spontaneous, often seemingly purposeful, formation of spatial, temporal, or spatiotemporal structures or functions in systems composed of a few or many components. In physics, chemistry and biology, self-organization occurs in open systems driven away from equilibrium.

Stochastic A stochastic or random process is a mathematical object usually defined as a collection of random variables. Stochastic fluctuations are significantly involved in the biological cellular environment. The simplest example is the Brownian motion whose peculiarities are an absence of memory and continuity.

Stoichiometry This is the calculation of the relative quantities of reactants and products involved in chemical reactions. It is founded on the law of conservation of mass, where the total mass of the reactants equals the total mass of the products, leading to the insight that the relations among quantities of reactants and products typically form a ratio of positive integers. Therefore, if the amounts of the separate reactants are known, then the amount of the product can be calculated.

Stochastic Thermodynamics It is born on the wake of the researches of Einstein, published in 1905, in which the century-old puzzle of the Brownian motion, was explained by showing that the erratic movements of small particles dispersed in a liquid was caused by the bouncing off from the atoms present in the fluid. More generally it concerns the behaviour of small systems embedded in an aqueous solution, as the colloidal particles, (bio) polymers as DNA and RNA, proteins, enzymes and molecular motors.

A stochastic process is usually described through a collection of random variables, whose fluctuations are present, for instance, in biological cells environment. The more common example is the motion of a Brownian particle, described through Eq. (1.3), whose peculiarity is the absence of memory. The applications in progress refer to the behaviour of molecular motors, where the transduction of chemical energy into mechanical work is determined by the constraints of the molecular architecture onto motion. Such a transduction can be formulated in terms of chemical and mechanical forces which induce directionality in the occuring movements. In this framework, for instance, the behaviour of the ATP synthase can be assumed to be a stochastic process since the arrivals and departures of each substrate molecule (ATP, ADP and phosphoric acid) in the catalytic sites present in the motor, are random events in time.

Probably the most important result of stochastic thermodynamics is the so called fluctuation theorem, introduced by Gavin Crooks. It is associated with the forward and reverse changes that a system, starting at the thermodynamic equilibrium with a thermal environment, undergoes as it is driven away by the action of a perturbation. After the perturbation ceases the system relaxes back to the equilibrium. During the non-equilibrium fluctuation, the probability ratio of the forward and the reversed conjugated trajectory is not unity, but it is instead given by the exponential of the total change in entropy along the forward process:

$$\frac{P(forward)}{P(reversed)} = \exp(\Delta S_{total})$$

In other world the overall entropy change present in the right hand side is a quantitative measure of the breaking of time-reversal symmetry at the level of trajectories.

Synthetic biology This is an interdisciplinary branch in which biology merges with engineering. It combines various disciplines from within different domains, such as biotechnology, genetic engineering, molecular biology, molecular engineering, systems biology, biophysics, electrical engineering, computer and control engineering, and evolutionary biology.

System biology This is the computational and mathematical approach to the modeling of complex biological systems. It is a biology-grounded interdisciplinary field of investigation that focuses on the deepening and simulation of the complex interactions present within biological systems.

Taxonomy The science or technique of classification of certain objects into ordered categories. In biology, it deals with the description, identification, naming, and classification of organisms.

Thermodynamics The name conferred, in 1854, by Lord Kelvin onto the mechanical theory of heat. A thermodynamic system consists of a certain amount of matter in the form of different chemical compounds, which occupies a particular region of space, separated from the external world, called the environment. It is assumed that the boundary conditions defining the energy and material exchanges that can take place between the system and the external world are defined at the separation surface. If such exchanges cannot take place, the system is defined as isolated.

Transcription This is the first step in gene expression. It involves the copy of a sequence of DNA of a gene for the purpose of making an RNA molecule. It is performed by enzymes called RNA polymerase, which link nucleotides to form an RNA strand, using it as a template. The whole process occurs in three stages: initiation, elongation and termination.

Vitalism This is the belief that living organisms are fundamentally different from non-living entities, because they contain some non-physical element or are governed by different principles than inanimate things.

Sources

General

Matthew Cobb. *Life's Greatest Secret*, PROFILE BOOKS, London 2015.
Harold Franklin. *The Way of the Cell*, Oxford University Press, 2001.
Falkowski, Paul G., *Life's Engines*, Princeton University Press, 2015.
Hoffmann, Peter. *Life's Ratchet*, Basic Books, New York, 2012.
De Duve, Christian. *Singularities*, Cambridge, 2005.
Reif Frederick. *Fundamental of statistical physics*, McGraw-Hill, New York, 1965.
Mayfield John E. *The Engine of Complexity*, Columbia University Press, 2013.
Bray Dennis. *Wetware,* Yale University Press, 2009.
Haynie Doald T. *Biological Thermodynamics*, Cambridge, August 2006.
Lane Nick. *The vital question*, Profile Books, 2015.
Dawkins Richard. *The ancestor's tale: A pilgrimage to the dawn of evolution*, WEINDELFELD and NICOLSON, London, 2004.
Judson Horace Freeland. *The Eighth Day of Creation*, Cold Spring Harbor Laboratory Press, 1966.
Hazen, Robert. *Genesis*, National Academy Press, 2005.
Luisi P. L., *The Emergence of Life*: From *Chemical Origins to Synthetic Biology*, Cambridge University Press, 2006.
Gavin E. Crooks, On Thermodynamic microscopic reversibility, Journal of statistical thermodynamics: Theory and Experiment, 2011 https://doi.org/10.1088/1742-5468/2011/07/P070083
Sara Imari Walker, Origins of Life: a problem for physics, a key issue review, Report progress in Physics, 2017 https://doi.org/10.1088/1361-6633/aa7804

© Springer Nature Switzerland AG 2018
S. Carrà, *Stepping Stones to Synthetic Biology*, The Frontiers Collection,
https://doi.org/10.1007/978-3-319-95459-2

Titles in This Series

Quantum Mechanics and Gravity
By Mendel Sachs

Quantum-Classical Correspondence
Dynamical Quantization and the Classical Limit
By A.O. Bolivar

Knowledge and the World: Challenges Beyond the Science Wars
Ed. by M. Carrier, J. Roggenhofer, G. Küppers and P. Blanchard

Quantum-Classical Analogies
By Daniela Dragoman and Mircea Dragoman

Quo Vadis Quantum Mechanics?
Ed. by Avshalom C. Elitzur, Shahar Dolev and Nancy Kolenda

Information and Its Role in Nature
By Juan G. Roederer

Extreme Events in Nature and Society
Ed. by Sergio Albeverio, Volker Jentsch and Holger Kantz

The Thermodynamic Machinery of Life
By Michal Kurzynski

© Springer Nature Switzerland AG 2018
S. Carrà, *Stepping Stones to Synthetic Biology*, The Frontiers Collection,
https://doi.org/10.1007/978-3-319-95459-2

Weak Links
The Universal Key to the Stability of Networks and Complex Systems
By Csermely Peter

The Emerging Physics of Consciousness
Ed. by Jack A. Tuszynski

Quantum Mechanics at the Crossroads
New Perspectives from History, Philosophy and Physics
Ed. by James Evans and Alan S. Thorndike

Mind, Matter and the Implicate Order
By Paavo T.I. Pylkkanen

Particle Metaphysics
A Critical Account of Subatomic Reality
By Brigitte Falkenburg

The Physical Basis of the Direction of Time
By H. Dieter Zeh

Asymmetry: The Foundation of Information
By Scott J. Muller

Decoherence
and the Quantum-To-Classical Transition
By Maximilian A. Schlosshauer

The Nonlinear Universe
Chaos, Emergence, Life
By Alwyn C. Scott

Quantum Superposition
Counterintuitive Consequences of Coherence, Entanglement, and Interference
By Mark P. Silverman

Symmetry Rules
How Science and Nature are Founded on Symmetry
By Joseph Rosen

Mind, Matter and Quantum Mechanics
By Henry P. Stapp

Entanglement, Information, and the Interpretation of Quantum Mechanics
By Gregg Jaeger

Relativity and the Nature of Spacetime
By Vesselin Petkov

The Biological Evolution of Religious Mind and Behavior
Ed. by Eckart Voland and Wulf Schiefenhövel

Homo Novus-A Human without Illusions
Ed. by Ulrich J. Frey, Charlotte Störmer and Kai P. Willführ

Brain-Computer Interfaces
Revolutionizing Human-Computer Interaction
Ed. by Bernhard Graimann, Brendan Allison and Gert Pfurtscheller

Extreme States of Matter
On Earth and in the Cosmos
By Vladimir E. Fortov

Searching for Extraterrestrial Intelligence
SETI Past, Present, and Future
Ed. by H. Paul Shuch

Essential Building Blocks of Human Nature
Ed. by Ulrich J. Frey, Charlotte Störmer and Kai P. Willführ

Mindful Universe
Quantum Mechanics and the Participating Observer
By Henry P. Stapp

Principles of Evolution
From the Planck Epoch to Complex Multicellular Life
Ed. by Hildegard Meyer-Ortmanns and Stefan Thurner

The Second Law of Economics
Energy, Entropy, and the Origins of Wealth
By Reiner Kümmel

States of Consciousness
Experimental Insights into Meditation, Waking, Sleep and Dreams
Ed. by Dean Cvetkovic and Irena Cosic

Elegance and Enigma
The Quantum Interviews
Ed. by Maximilian Schlosshauer

Humans on Earth
From Origins to Possible Futures
By Filipe Duarte Santos

Evolution 2.0
Implications of Darwinism in Philosophy and the Social and Natural Sciences
Ed. by Martin Brinkworth and Friedel Weinert

Probability in Physics
Ed. by Yemima Ben-Menahem and Meir Hemmo

Chips 2020
A Guide to the Future of Nanoelectronics
Ed. by Bernd Hoefflinger

From the Web to the Grid and Beyond
Computing Paradigms Driven by High-Energy Physics
Ed. by René Brun, Frederico Carminati and Giuliana Galli-Carminati

The Language Phenomenon
Human Communication from Milliseconds to Millennia
Ed. by P.-M. Binder and K. Smith

The Dual Nature of Life
Interplay of the Individual and the Genome
By Gennadiy Zhegunov

Natural Fabrications
Science, Emergence and Consciousness
By William Seager

Ultimate Horizons
Probing the Limits of the Universe
By Helmut Satz

Physics, Nature and Society
A Guide to Order and Complexity in Our World
By Joaquín Marro

Extraterrestrial Altruism
Evolution and Ethics in the Cosmos
Ed. by Douglas A. Vakoch

The Beginning and the End
The Meaning of Life in a Cosmological Perspective
By Clément Vidal

A Brief History of String Theory
From Dual Models to M-Theory
By Dean Rickles

Singularity Hypotheses
A Scientific and Philosophical Assessment
Ed. by Amnon H. Eden, James H. Moor, Johnny H. Søraker and Eric Steinhart

Why More Is Different
Philosophical Issues in Condensed Matter Physics and Complex Systems
Ed. by Brigitte Falkenburg and Margaret Morrison

Questioning the Foundations of Physics
Which of Our Fundamental Assumptions Are Wrong?
Ed. by Anthony Aguirre, Brendan Foster and Zeeya Merali

It From Bit or Bit From It?
On Physics and Information
Ed. by Anthony Aguirre, Brendan Foster and Zeeya Merali

Computer Simulations in Science and Engineering
Concepts – Practices – Perspectives
By Juan M. Durán

Stepping Stones to Synthetic Biology
By Sergio Carrà

Printed in the United States
By Bookmasters